The Aesthetics of Science

This volume builds on two recent developments in philosophy on the relationship between art and science: the notion of representation and the role of values in theory choice and the development of scientific theories. Its aim is to address questions regarding scientific creativity and imagination, the status of scientific performances—such as thought experiments and visual aids—and the role of aesthetic considerations in the context of discovery and justification of scientific theories.

Several contributions focus on the concept of beauty as employed by practising scientists, the aesthetic factors at play in science and their role in decision making. Other essays address the question of scientific creativity and how aesthetic judgment resolves the problem of theory choice by employing aesthetic criteria and incorporating insights from both objectivism and subjectivism. The volume also features original perspectives on the role of the sublime in science and sheds light on the empirical work studying the experience of the sublime in science and its relation to the experience of understanding.

The Aesthetics of Science tackles these topics from a variety of novel and thought-provoking angles. It will be of interest to researchers and advanced students in the philosophy of science and aesthetics, as well as other subdisciplines such as epistemology and philosophy of mathematics.

Steven French is Professor of Philosophy of Science at the University of Leeds, UK. He is a co-editor of *Thinking about Science, Reflecting on Art* (Routledge, 2017) and Co-Editor-in-Chief of *The British Journal for the Philosophy of Science*.

Milena Ivanova is a teaching associate in the Department for the History and Philosophy of Science and a bye-fellow at Fitzwilliam at the University of Cambridge, UK.

Routledge Studies in the Philosophy of Science

Science after the Practice Turn in Philosophy, History, and the Social Studies of Science
Edited by Léna Soler, Sjoerd Zwart, Vincent Israel-Jost, and Michael Lynch

Causation, Evidence, and Inference
Julian Reiss

Conceptual Change and the Philosophy of Science
Alternative Interpretations of the A Priori
David J. Stump

Neo-Aristotelian Perspectives on Contemporary Science
Edited by William M.R. Simpson, Robert C. Koons and Nicholas J. Teh

Essence in the Age of Evolution
A New Theory of Natural Kinds
Christopher J. Austin

The Instrument of Science
Scientific Anti-Realism Revitalised
Darrell P. Rowbottom

Understanding Perspectivism
Scientific Challenges and Methodological Prospects
Edited by Michela Massimi and Casey D. McCoy

The Aesthetics of Science
Beauty, Imagination and Understanding
Edited by Milena Ivanova and Steven French

For more information about this series, please visit: www.routledge.com/Routledge-Studies-in-the-Philosophy-of-Science/book-series/POS

The Aesthetics of Science

Beauty, Imagination and Understanding

Edited by Milena Ivanova and Steven French

Routledge
Taylor & Francis Group
LONDON AND NEW YORK

First published 2020 by Routledge

2 Park Square, Milton Park, Abingdon, Oxon OX14 4RN
605 Third Avenue, New York, NY 10017

Routledge is an imprint of the Taylor & Francis Group, an informa business

First issued in paperback 2022

Copyright © 2020 Taylor & Francis

The right of the editors to be identified as the authors of the editorial material, and of the authors for their individual chapters, has been asserted in accordance with sections 77 and 78 of the Copyright, Designs and Patents Act 1988.

All rights reserved. No part of this book may be reprinted or reproduced or utilized in any form or by any electronic, mechanical, or other means, now known or hereafter invented, including photocopying and recording, or in any information storage or retrieval system, without permission in writing from the publishers.

Notice:
Product or corporate names may be trademarks or registered trademarks, and are used only for identification and explanation without intent to infringe.

Publisher's Note

The publisher has gone to great lengths to ensure the quality of this reprint but points out that some imperfections in the original copies may be apparent.

Library of Congress Cataloging-in-Publication Data
Names: French, Steven, editor. | Ivanova, Milena, editor.
Title: The aesthetics of science : beauty, imagination and understanding / edited by Steven French and Milena Ivanova.
Description: New York : Routledge, 2020. | Series: Routledge studies in the philosophy of science | Includes bibliographical references and index.
Identifiers: LCCN 2019051567 (print) | LCCN 2019051568 (ebook) | ISBN 9780367141141 (hardback) | ISBN 9780429030284 (ebook)
Subjects: LCSH: Art and science. | Science—Philosophy.
Classification: LCC N72.S3 A325 2020 (print) | LCC N72.S3 (ebook) | DDC 701/.05—dc23
LC record available at https://lccn.loc.gov/2019051567
LC ebook record available at https://lccn.loc.gov/2019051568

ISBN: 978-0-367-14114-1 (hbk)
ISBN: 978-1-03-233718-0 (pbk)
DOI: 10.4324/9780429030284

Typeset in Sabon
by Apex CoVantage, LLC

Contents

Preface

This volume is the end result of a conversation that began at a conference in Konstanz in January 2016 and was continued later that year at the annual conference of the British Society for the Philosophy of Science in Cardiff. With funding from the British Society of Aesthetics and the British Society for the Philosophy of Science and with the assistance of Alice Murphy, we organised a conference on the 'Aesthetics of Science' at the University of Leeds in July 2017, featuring a number of the contributors to this collection. This wasn't the first such conference on this topic by any means—indeed, there were two previous workshops in this area at Leeds alone—but we would argue that it was one of the most agreeable, certainly of those we have attended! The presentations were uniformly excellent, the discussions were insightful and engaging, and the atmosphere overall was constructive and supportive. We hope that some of those qualities have carried over to the chapters collected here that represent a diversity of perspectives and approaches but also display a number of common themes and concerns. In our 'Introduction' we try to set those issues out as clearly as we can and highlight not only the commonalities but also the interesting distinctions and divergences going forward. And we hope that you, the reader, get as much pleasure and insight out of these pieces as we did reading them and putting the volume together!

Milena and Steven

Acknowledgments

As well as the financial support of the BSA and the BSPS and the logistical help of Alice Murphy, we'd also like to thank all the contributors for their willingness to participate and, particularly in the final stages, for their patience.

Milena would also like to thank Matt and their daughter Cailyn, who arrived into the world during the development of this project.

Steven would like to thank, as always, Dena and Morgan, and a certain Ruffian.

1 Introduction to the Volume

Milena Ivanova and Steven French

1 Introduction

Aesthetic judgments feature prominently in scientific practice. Scientific theories are often compared to works of art, with scientists likening the process of constructing a theory to that of creating art pieces and even in choosing one theory over another they may invoke aesthetic considerations. Given these features of scientific practice, the questions naturally arise: What are the inter-relationships between aesthetics and science? How can the role of aesthetic judgments in scientific practice be justified? This volume engages with these questions and considers in detail the status of various features of such practice from an aesthetics-related perspective, including thought experiments and models, visual aids and representations, together with the role of aesthetic considerations in the context of discovery and justification of theories, the experiences of beauty and the sublime in science and how they affect and shape scientific practice, and the nature of scientific creativity and imagination in general.

2 History of Engagement

Engaging in the *aesthetics* of science has certainly not always been a topic of pursuit in the philosophy of science. During the positivist dominated years aesthetics and science were kept apart, and there was little value seen in the engagement between the two disciplines. For one, aesthetic considerations, if indeed relevant to science, were deemed to be psychological and subjective in nature, and though they might be employed in the process by which scientists come up with ideas, they were regarded to have no bearing upon the formal properties of the theory, that is, how the theory relates to the world. Hans Reichenbach's famous distinction between the context of discovery and the context of justification conveys exactly this point: it is irrelevant how scientists come up with new ideas, whether they dream them up in their sleep or have a sudden illumination whilst taking a stroll, what matters is whether the reasons used can justify one's belief that the theory corresponds to the world. Reichenbach claimed that "It would be a vain attempt to construct a theory of knowledge which is at the same time logically complete and in strict correspondence with the psychological process of thought" (1938: 5). Thus, from the time of the Vienna Circle through to recent years, aesthetic considerations were not the focus of philosophical work. If acknowledged at all as featuring in the practice of science, such considerations were clearly demoted to the context of discovery, rendering them not part of the rational justification of scientific theories.

3 The Significance of Representation

Certain philosophical developments over the last decades have paved the way for departing from the constraints of Reichenbach's distinction and engaging systematically with the relationship between aesthetics and science. One contributing factor has been the popularity of the semantic approach to scientific theories and the increased interest in the function of scientific models. While the syntactic approach that dominated the years of logical positivism took theories to be sets of propositions that are truth apt, the alternative semantic, or model-theoretic approach, as famously outlined by Patrick Suppes (1960), introduced the notion of representation as the aim of theories. How theories represent the world became the central question, with many commentators drawing analogies with the representational nature of artworks and scientific theories explicitly compared to such artworks. Bas van Fraassen's (2008) seminal work *Scientific Representation* engaged systematically with the notion of representation in art and in science, and the edited collection by Frigg and Hunter (2010) *From Mimesis to Representation*, further explored the relationship between scientific models and works of fiction. The recent volume *Thinking about Science, Reflecting on Art: Bringing Aesthetics and Philosophy of Science Together* (2017), edited by Bueno et al., gave additional momentum to this engagement, opening further avenues for exploration such as the act of interpretation in art and science and the question of whether there can be a science of aesthetics. Further connections between art and science were also introduced in the work of Catherine Elgin (1991) and others, who compared literary works to thought experiments, showing how our understanding can be advanced through notions such as exemplification, for example.

4 Beauty in Science

Beyond the notion of representation and the comparison of scientific products such as theories and thought experiments to artworks, philosophers of science have also focused on the notion of beauty itself. The work of James McAllister, *Beauty and Revolutions in Science*, set the stage for understanding the notion of beauty within the historical evolution of scientific theories. It offered an account of how scientists come to form aesthetic judgments and how their training affects their aesthetic appreciation. McAllister also provided a justification for the idea that aesthetic considerations can play an epistemic role. While many scientific realists, contemporary and past, have tried to identify the theoretical virtues that correlate with the truthlikeness of theories, McAllister and others after him explicitly recognised that these virtues are often conveyed in aesthetic terms and noted that scientists explicitly use aesthetic language when they appraise them, recognising the need to give an account of the aesthetic aspect of these judgments. Recent developments have seen a renewed appreciation for the role of certain values in theory choice and the development of scientific theories, as exemplified in Samuel Schindler's (2018) *Theoretical Virtues in Science: Uncovering Reality Through Theory*. Furthermore, there is recognition that when scientists engage with theories that they consider beautiful they are indeed reporting genuine aesthetic experiences (Ivanova 2017a).

5 Science and Creativity

In addition to these emphases on representation and aesthetic qualities, historians of science, psychologists and neuroscientists have become invested in understanding the notion of creativity. Historians try to understand how scientists of the past came up with the new theories that revolutionalised their fields, psychologists try to understand what traits creative people have in common and how such traits are formed, while neuroscientists have focused on understanding the neurological functions involved in the exhibition of creative behaviour. The departure from the 'inspirationalist' accounts of creativity, which deemed inspiration to be a mysterious process available only to a select set of individuals, the 'great minds', has opened the door to the exploration of the creativity and the imagination in terms of computation, as advanced in the work of Margaret Boden. Here again the connection between art and science has become apparent, with creativity being highly valued in both the domain of art and the domain of science.

6 From Aesthetics to Philosophy of Science (and Back Again)

This volume extends this increased engagement between aesthetics and science of recent years and introduces new avenues for exploration. The collection focuses on the status of aesthetic judgments with regard to the products of science, the status of scientific theories seen as constructions of scientific imagination, the experience of beauty but also the sublime in science, how aesthetic considerations inform and shape our activities and aims in science and, finally, the question of scientific creativity. There are important dimensions to science practice whose nature departs from the logical positivist's recipe of 'logic and experience', both in the context of discovery and justification, and entering the field of aesthetics, that need to be systematically explored. For one, scientists often make explicit aesthetic judgments with regard to the objects they study, the products of their activities as well as those very activities seem to be guided by aesthetic values. The phenomena studied in science are claimed to be beautiful, such as the diffraction of light rays or solar eclipses. More significantly, we find claims that the products of scientists' activities themselves exemplify aesthetic values, with physicists typically claiming theories such as Einstein's relativity theory or Newton's mechanics to be beautiful, Rutherford's experiments on the atom to be beautiful, Watson, Crick and Franklin's double helix model of DNA molecules to be beautiful and so on. And the very construction of a theory or an experiment can be claimed to be guided by aesthetic considerations. Since aesthetic judgments enter in all these levels of theorising, there is a need to understand the nature of these aesthetic judgments and the role they play. What are the set of aesthetic judgments that guide scientists? Are they fixed once and for all, and across disciplines, or are they largely contingent, relevant to a framework, school of thought and time period?

Debates in aesthetics have aimed to resolve the very same question when it comes to artworks. According to objectivism, aesthetic judgments have validity across individuals, time frameworks and societies, meaning that there is a fact of the matter whether a certain object is beautiful or not. Objectivists argue that aesthetic judgments can be regarded as independent of subjective taste and fashions and point to works of art that have continuously been appreciated cross culturally and through time. For instance, we value the works of Callicrates, Polykleitos and Homer today as they were valued in antiquity, supporting the idea that our aesthetic judgments are objective and do not change with time or across societies. On the other hand, some artworks can initially be regarded as 'ugly' or aesthetically displeasing, but gain ground later, suggesting that aesthetic judgments can be subjective, contingent and varying across time, communities and individuals. The infamous reception of the Eiffel tower exemplifies this point. Most artists and architects in the nineteenth century wanted the tower demolished, calling it a 'monstrosity' over the Parisian skyline, but only a decade later the tower became a symbol of modern architecture and regarded as one of the most beautiful buildings in the world. Similarly in science, some values seem to gain ground after they are introduced in the scientific community. For instance, symmetry was not praised before relativity theory, elegance was irrelevant before the mathematical formalisation of theories, culminating with Newton's development of the calculus and his theory of gravity. What shapes the community's response to aesthetic qualities of theories and what roles these can play are questions that are beginning to receive more systematic attention in the contemporary literature.

Another aspect of productive engagement between aesthetics and philosophy of science concerns creativity and the imagination. We value original ideas and the creative process responsible for their generation. When it comes to artworks, we do not ascribe value to copies or forgeries; we value originals. In science, we praise and admire those who discover new theories, phenomena and design new experiments or instruments, those who produce new proofs rather than those scientists who replicate experiments, or come up with a theory second or third. The reward structure in science reflects this phenomenon; credit attribution goes to those who discover first. We grant Nobel Prizes for new discoveries, while not much value is given to those who replicate experiments, for instance, leading to problems such as the replication crisis (Heesen 2018). As the sociologist Robert Merton reflected, science is governed by the priority rule, the fight to be the first who comes up with new ideas. How do artists and scientists do this?

Galileo, Newton, Curie, Einstein and Poincare are the usual examples given of creative minds, geniuses raised to the status of mythical superheroes endowed with creativity and imagination that transformed the field and our understanding of the world. But was there anything special about these scientists? Creativity has been the focus of much attention in aesthetics. Earlier 'inspirationalist' accounts took creativity to be due to divine or special inspiration available to very few individuals, but more recently systematic work in psychology and neuroscience has illuminated the creative process and the social and cultural aspects that enable some individuals to develop creativity. In the work of Boden (2003) and others creativity is understood as the exploration of conceptual spaces and the ability to connect already known ideas, with value being ascribed only to those connections that are historically novel. Within this new way of thinking about creativity, interesting questions arise, such as whether creative individuals share the same traits, how creativity can be cultivated, and how an individual's environment, social and cultural background and resources available to them can affect that creativity. This also generates questions regarding credit distribution and recognition that creativity could be explored from the perspective of groups rather than individuals (Currie 2019). Interesting new dimensions in the study of creativity has also recently been raised in the work of virtue epistemology, where creativity is construed as an epistemic virtue whose instantiation in an individual leads to epistemic success. A troubling issue for virtue epistemology is to reconcile the descriptive and normative aspect: as a matter of fact, biographical accounts often reveal that creative people exemplify a lot of epistemic vice, from self-centeredness, dogmatism and ego-centric bias, to egotism and narcissism. How are we to reconcile the idea of the virtuous knower with the descriptive aspect that new revolutionary ideas that lead to scientific progress and epistemic success are a product of epistemic vice? The problem opens the door to reconsidering the notion of creativity within both virtue and social epistemology and exploring the creative process from the dimension of groups and individuals.

The new engagement between philosophers of science and aesthetics has also motivated work in the history of philosophy of science and the search for ideas that predate the logical positivist distinction between the contexts of discovery and justification. Here philosophers have uncovered interesting work on aesthetic aspects of science developed before or during the logical positivist movement, from Dirac's arguments on beauty by Graham Farmelo, to Praisly Livingston's revival of Poincare's sophisticated account of creativity and scientific discovery, to David Stump's revival of Pierre Duhem's use of 'good sense' in theory choice and Milena Ivanova's recent reconstruction of Poincare's account of beauty in science. These works show that there was a systematic engagement in the early twentieth century with the notion of creativity in the context of scientific discovery, the notion of beauty as a guide and evaluator in scientific reasoning, and the role of aesthetic sensibility in scientific decision making, all of which can be productively reintroduced into our contemporary engagement in this field.

As two of the contributors to this volume, Arcangeli and Dokic, note, '[a]esthetics seems to enter science on at least three different levels:

(i) The *objects* of scientific enquiry (such as cells, mu-mesons, and numbers) may instantiate aesthetic values.
(ii) The *products* of science (such as theories, conjectures, and models) may instantiate aesthetic values.
(iii) The scientific *practice* (such as constructing and evaluating theories, and designing experiments) may be guided by aesthetic experiences and judgements.'

The contributions in this book focus primarily on (ii) and (iii), the products or 'outputs' of science, including not just theories and models but also thought experiments, for example, and the practices, covering, in addition to theory discovery and justification, the presentation of theories at lectures and seminars. We'll also look at the practitioners of science, not just in terms of what they do and produce but the virtues and vices that they exhibit. In doing so we shall address various aspects of the above issues from a variety of perspectives that, we hope, will further advance the engagement between aesthetics and philosophy of science in general.

7 Summary of Contributions

In the opening contribution to the volume, Catherine Z. Elgin addresses head-on 'the problem of the aesthetic' in the context of science: is there any epistemically good reason to prefer a theory that possesses certain aesthetic qualities to one that does not? And what are we focusing on when we make such assessments? Extending a view found in the philosophy of art, Elgin suggests that aesthetic responses to theories consist in the apprehension and appreciation of 'scientifically significant forms in a logical space', where the nature of these forms is context dependent. Furthermore, she argues, the role of the relevant aesthetic factors is not merely instrumental nor is it truth-conducive; rather these factors act as 'gatekeepers' to the acceptability of theories. In her earlier work she has developed the view that an understanding of a given topic consists of a systematically linked body of information in reflective equilibrium that is grounded in fact, is responsive to evidence and enables non-trivial inference and argument about a range of phenomena. And insofar as an aesthetic factor is ineliminably integral to such a network of scientific commitments, then it is epistemically justified.

Thus she considers the role of symmetry in modern science, regarded as an aesthetically pleasing feature. The recent history of physics demonstrates how scientists prefer symmetry-preserving theories and this preference clearly affects their behaviour in accepting or rejecting new hypotheses or results in general. Another factor is systematicity; as she puts it, '[w]e want our fabric of scientific commitments to be tightly woven'. Yet another is simplicity, notoriously complicated as she points out. Different kinds of simplicity may be traded off against one another and come to the fore in different contexts. Nevertheless, construing it as an aesthetic factor helps explain scientists' preferences for simpler theories and models.

As she goes on to note, such aesthetic considerations may be initially tenable and thereby constrain future theorising. Candidate theories that display these qualities will be deemed to be acceptable over those that do not. However, every component of that systematic body of information in equilibrium is up for grabs and it may of course turn out that the cost of insisting on a particular quality is too high, so that its scope must be reduced or it is abandoned altogether. Conversely, a particular factor may gain in importance, as in the case of symmetry in the shift from classical to quantum physics. Thus aesthetic factors play a regulatory role, helping to shape our accounts and frame our understanding.

The way in which the aesthetic features of scientific theories and models contribute to the understanding associated with them is one of the themes running through a number of the contributions. In her 'Getting the picture: Towards a new account of scientific understanding', Letitia Meynell connects recent work on the nature of scientific understanding with these aesthetic features of science by advancing her 'pictorial' account of the former. According to this, the characteristic content of understanding is pictorial rather than propositional and it is by virtue of the epistemic flexibility of pictures that, Meynell claims, her account can embrace unificatory, mechanistic and pluralist views of understanding. Two features of pictures are key to their exemplification of the cognitive processes and values characteristic of understanding: first, their affinity to visual experiences via their two-dimensional character. It is through this feature that the unificatory aspect of understanding can be accommodated, as the relevant aspects under study are brought together into a whole by the viewer. And insofar as there will be a range of legitimate ways of reading an image, this account can also capture the way in which understanding comes in degrees.

The second feature concerns their representational role, in which they serve as props for the imagination. In this respect de Regt's emphasis on visualisation can be accommodated, but so can Woodward's and Salmon's causal theory, since the causal basis of our visual experience of reality will be carried over to our comprehension of pictures that represent the relevant states of affairs through their spatial features.

Meynell further draws on Walton's work in aesthetics to argue that different people may possess different understandings of, say, a genetic ribbon diagram by virtue of bringing to bear a different array of beliefs, habits of mind, conventions and so forth—that Walton calls 'principles of generation'—that constrain but do not determine what is to be imagined and under what circumstances. Given the distinction between the system under study or the artwork and the viewer's experience of it, this can then account for how different viewers may have different understandings of the same work, which further illuminates the differences between subjective and inter-subjective understanding in science.

To illustrate her account, Meynell draws on her previous work on Feynman diagrams, arguing that they help to unify the relevant phenomena in a cognitively accessible way, display complex causal connections within those phenomena and allow scientists to bring their own interests and commitments to bear on the interpretation of quantum field theory. She also considers the understanding required to give informed consent in research, arguing that this involves not only the acquisition of relevant information but the integration of various different types in a way that accommodates how participation may cause various complications for the participant as well as the positive outcomes. As she notes, this can be compared with Elgin and Goodman's 'world-making' and although this isn't the same as understanding the research itself it does involve at least a rough sketch of the science in question and some grasp of what participation in the research will imply. Through such examples Meynell elegantly weaves together themes from aesthetics and philosophy of science to lay down the basis of a more nuanced account of understanding across both domains.

Todd also considers the role of imagery in science but adopts a different tack. Focusing on imagistic imagination or 'visualisation' in thought experiments and scientific models, he rejects recent 'fictionalist' accounts of these outputs of scientific activity that draw on Walton's work. The problem he finds with the Waltonian account of fictional engagement is that for thought experiments and models to play the roles that they do in scientific practice, certain constraints need to be applied, yet the account is notably lacking in detail as to the nature and origin of these constraints. Ultimately, he insists, we must look elsewhere if we are to shed any light on the cognitive value of thought experiments and models and on the epistemic function of imagination that is involved.

Underpinning such accounts, he suggests, is a view of imagery that takes it to be epistemically useless. And this in turn is grounded in the idea that imagery is 'transparent', in the sense that although we may employ images to help us imagine, we see through them the things we are actually imagining, as it were. However, he argues, when we have an imaginative experience, there are certain phenomenological and structural features of that experience that we are aware of, simply by virtue of having it. Indeed, it is the relative 'opacity' of imagery that leads it to play a significant cognitive role in scientific reasoning.

Indeed, he goes on to suggest, imagery might possess important cognitive value arising from its connection with certain affective states that themselves possess cognitive import. The imaginative contemplation of certain thought experiments may evoke certain quasi-sensory intuitions on the basis of which new beliefs can be formed. More generally, there are broader connections between imagery and affect that support a more expanded role to imagery when it comes to scientific models. In such cases, certain salient aspects are highlighted, and certain patterns recognised through the engagement with imagistic imaginings and this derives from the deep connection between imagery and aesthetic feelings and the epistemic function that the latter serve in scientific reasoning.

Two such are the 'feeling of knowing' that has to do with the accessibility of the knowledge that one has, and the 'feeling of understanding' that concerns the intellectual satisfaction that motivates the endorsement of a scientific explanation. As Todd notes, there exists evidence for neural correlates of these feelings, and also support for the claim that some epistemic feelings do play a justificatory role in accurately predicting future cognitive performance, as well as in acting as a stimulus to judgment. These feelings of knowledge and understanding play a central role in scientific reasoning and in the development and application of scientific models but, as Todd has argued elsewhere, we can also regard them as possessing aesthetic attributes. Thus they are 'valenced', whether positively or negatively, they are typically 'quick and dirty' responses that are opaque in certain ways and, crucially, they are often characterised as aesthetic in scientific practice. Furthermore, when it comes to 'understanding' and 'fit' we find striking continuities between aesthetic judgments and scientific ones, in terms of the appreciation of patterns, connections, symmetries and harmonies in each case. The kind of understanding that is involved here, Todd argues, is not primarily propositional—it is imagistic, affective and, crucially, typically has an aesthetic character.

As he says in conclusion, in those cases in scientific practice for which visualisation and imagery are unavoidable, they are typically accompanied by feelings that arise from the phenomenal character of the imagining itself. And the feeling associated with understanding or 'fit' in such cases is not that which could be associated with some passive perceptual or belief state. Rather, there is conscious effort involved in such cases that manifests in aesthetic-epistemic feelings tied to the kinds of features that are examined from different perspectives in this volume; that is, features such as harmony, unity, symmetry and so forth.

Understanding also features prominently in Ivanova's contribution, where she argues against accounts that seek to relate the aesthetic qualities of a scientific theory to its likelihood of being *true* and maintains, instead, that aesthetic factors are tightly bound up with our own cognitive make up and the desire to understand the world around us. As she notes, beauty, whether exemplified by the elegance of a theory, or its simplicity, or its unity, not only functions as a heuristic factor in theory discovery and pursuit, but is often regarded as an indicator of truth. Scientists themselves, from Poincaré to Dirac have appealed to beauty when justifying their commitment to a theory. However, she asks, what is the connection here? On the one hand, it might be suggested that beautiful theories correctly capture facts about the world, so their aesthetic qualities reflect the beauty in the world. But this, of course, runs up against objections that it assumes that the world is beautiful, in whatever sense, and modern physics, as embodied in the Standard Model of elementary particle physics might suggest otherwise.

On the other hand, it could be argued that we should have confidence in a beautiful theory simply because, inductively, beautiful theories have had a good track record of empirical success. Here she draws on McAllister's use of the 'exposure effect' from psychology: scientists learn from exposure to the aesthetic qualities of past successful theories what features to invoke in the pursuit and evaluation of current theories. We can then induce that future theories possessing these qualities will be successful. Furthermore, the language in which such features are described should be taken literally, as discourse about the aesthetic qualities of the theories concerned, rather than a 'stand in' for non-aesthetic features.

However, Ivanova argues, this latter account cannot explain why certain qualities, such as simplicity and unity, persist throughout the history of science, while others fade from the scene, despite being associated with empirically successful theories. More significantly, she regards this sort of argument as unduly 'optimistic' and points out that as the realism debate illustrates, the history of science can be used to give a very different conclusion. There are, after all, beautiful theories that have failed and 'ugly' ones—such as quantum mechanics perhaps—that are hugely successful.

However, she suggests, if we shift our epistemic aim from truth to understanding, the regulative role played by these qualities can be recognised and appropriately accommodated within our philosophy of science. Appealing to non-factive accounts of understanding, such as Elgin's or de Regt's, that treat it as a skill or ability, then opens up epistemic space for aesthetic qualities to play a role. The elegance or unity of a theory then reflects, not certain features of the world, but our choice to construct it along these lines and that is because we are then better able to manipulate and work with it. Taking these qualities as conditions of our cognitive make-up then explains why certain of them persist even when our best theories fail to exhibit them—in such cases the qualities are shifted to potential future theories that are sought after. Thus, she concludes, the significance of such aesthetic qualities in scientists' decision-making has to do with the way we think about the world and is independent of the question whether such qualities can lead us to expect that the theories we regard as beautiful are also likely to be successful.

Margherita Arcangeli and Jérome Dokic broaden the discussion by focusing on the 'sublime' in their paper 'A Plea for the Sublime in Science'. The sublime is often taken to contrast with beauty, not least because the former can be disturbing as well as enlightening. They begin by considering how experiences of beauty and the sublime are contrastive in this way, with the latter manifesting an overwhelming and, often, negative aspect not present in the former. Beauty experiences tend to be positive and pleasurable, involving reward and satisfaction. Sublimity experiences, on the other hand, elicit a sense of vastness and grandeur that can be unsettling, at the very least. Both, however, can be considered aesthetic experiences, albeit corresponding to different cognitive patterns. And both play a role in science, although that played by the sublime has tended to be overlooked. On the one hand, the sublime can be an object of empirical investigation itself but on the other, it can also be a guide in scientific practice.

Thus, as Arcangeli and Dokic remark, scientists themselves use words and phrases such as 'mysterious', 'feeling of awe', 'frightening', that are evocative of sublimity experiences in both their positive and negative aspects. As they go on to describe, recent work by psychologists demonstrates that such experiences can themselves be the subject of scientific enquiry. Indeed, certain neurological findings indicate that experiences of the sublime activate different areas of the brain than do beauty experiences and also reflect the sense of a loss of self that are inherent in the former.

Furthermore, just as aesthetic qualities such as beauty can play an important role in scientific practice, so can the sublime. Indeed, a focus on the latter reveals significant features of such practice that might otherwise be overlooked. Consider, again, the relationship between truth and beauty. Whether this relationship can be ontologically grounded or not, it has been claimed that judgments about both share a common characteristic by virtue of being made through people attending to the fluency of their own information processing. Thus certain shapes are judged to be more beautiful than others because the relevant features can be processed more easily. Likewise, sentences that are more easily processed tend to be judged to be true. Interiorising the associated heuristic relating beauty and truth underpins the mechanism that links the fluency-based approach with the claim that such judgments play a role in scientific practice.

Significantly, this overall approach can then be extended along two avenues: first, to cover judgments of understanding and second, to include experiences of the sublime. Here Arcangeli and Dokic draw on a distinction between perceptual and conceptual fluency and argue that just as certain paintings, say, may be regarded as visually 'disfluent' but conceptually fluent, so experiences of the sublime in science may involve a similar balance of fluency and disfluency. Thus they suggest that judgments of sublimity may draw the attention of scientists to highly challenging phenomena and domains of enquiry and that these judgments may then contribute to the evaluation of a theory as innovative or ground breaking.

Finally, they insist, the objects of sublimity experiences are more relational than those of beauty experiences and this renders the former apt to ground deep judgments of understanding, with regard to the limits of human cognition. They may be characterised as 'limit-experiences' in the sense that they involve the feeling that the theory under consideration has been pushed towards the limits of what we may cognitively encompass as human beings. Sublimity experiences, then, may manifest at the deepest or most foundational levels of scientific practices, when we contemplate theories like quantum mechanics and general relativity and as such, they are deserving of more comprehensive philosophical treatment.

Meynell, Ivanova, Arcangeli and Dokic primarily focus on theories as the bearers of the relevant aesthetic qualities. However, other products of science can also possess them. Alexander Bird looks at explanations in his *How Can Loveliness be a Guide to Truth? Inference to the Best Explanation and Exemplars*. He begins by noting that scientists frequently invoke 'Inference to the Best Explanation' in order to assess the plausibility of their theories. Lipton famously claimed that in this process explanations are ranked according to their 'loveliness' with the top ranked explanation chosen as 'the best'. Bird then asks, what are the lovely making features of an explanatory theory? And how do scientists come to see and respond to these features in a theory? An immediate issue in canvassing answers to these questions has to do with the objection that loveliness is subjective. In that case, it cannot be held to correlate with the truth. In part, as Bird notes, this has to do with the ineffable nature of loveliness—we are better at recognising it than we are at articulating what it is. Furthermore, there is considerable variation across different disciplines about what counts as lovely in an explanation. These are concerns that may also be applied to aesthetic features in general; so, for example, we may feel that Mozart's oboe quartet is beautiful without being able to articulate in terms of what properties it is beautiful. How might we assuage these worries?

Bird's response is to appeal to the idea that a community's standards of explanatory goodness are acquired in the process of scientific training and learning which uses certain *exemplars* (and here we might recall Elgin's work on their role). As he says, this notion of a scientific exemplar originates with Kuhn who emphasised the exemplary nature of certain solutions to scientific puzzles that then drive the processes of scientific cognition. It is the perceived similarity with the relevant exemplars that determines theory choice in science. Furthermore, scientists learn how to tackle the problems they are faced with through repeated exposure to and practice with these exemplars.

However, Bird argues, we can take this overall framework further and place it in a naturalistic and realist context in which we can account for our ability to latch onto the truth in virtue of possessing certain reliable modes of thinking. Crucially, it is through exposure to exemplars that scientists acquire their standards of explanatory goodness that is of 'loveliness'. As Bird notes, this really goes all the way back to Aristotle and the idea that we acquire our virtues not through learning certain rules but through a process of training and being exposed to examples of virtuous behaviour. Of course, it is due to the nature of that process that the judgments reached are not wholly a matter of rational, conscious deliberation. As a result the scientist, in making such a judgment, may not be able to fully articulate the reasons for doing so. And the factors involved will be dependent on the exemplars that the scientist was exposed to in their training.

Here we can see how the above worries might be dealt with. Indeed the ineffability of scientists' judgments about the loveliness of an explanation is entirely to be expected if the process of learning occurs through training with exemplars rather than the acquisition of explicit rules. And likewise, different scientific domains will invoke different exemplars and hence what we have across science are multiple sets of criteria of explanatory goodness. Indeed, we might even see a shift in such criteria within a given domain across time.

Nevertheless, as Bird goes on to note, this approach does not guarantee that the relevant exemplars will establish truth-conducive standards. The emphasis on certain analogies in medieval science is illustrative of that. To use a more recent example, the replication crisis in social psychology suggests that its exemplars are also not truth-conducive. Having said that, if the exemplars underpinning certain criteria are themselves true, then, Bird argues, those criteria can be conducive of the truth. Given all this, if success does indeed correlate with the truth and in discerning the explanatory features of exemplars scientists are, in fact, discerning properties that play some role in the success of theories, then we may conclude that Inference to the Best Explanation is reliable. More generally, Bird argues that this exemplar-based approach exhibits a number of advantages over McAllistair's so-called 'aesthetic induction', mentioned above, according to which scientists make aesthetic judgments about theories that play a role in theory acceptance. In particular, Bird claims, the latter assumes that such judgments can be distinguished from those made on the basis of empirical criteria, whereas he is sceptical of this and insists that the exemplar based approach makes no such distinction. Given the central role of explanation in the assessment of theories, he maintains, we cannot assess a theory on empirical grounds independently of assessing it in terms of its explanatory goodness.

As he says in the conclusion, this role of exemplars can also be identified in the world of art where certain instances of 'great art' are held up as embodying the aesthetic qualities of beauty, sublimity and so forth, thereby setting the standard by which other works can be taken to possess these qualities. As in art so in science, where certain theories, historical episodes, problem solutions etc. are held up as exemplary. The difference, of course, is that in the case of the latter, if we adopt a realist stance, we should also maintain that the exemplar based standards of explanatory loveliness yield a preference for theories that are more likely to be true.

In her contribution, 'The Literary Form of Scientific Thought Experiments' Alice Murphy extends this focus from theories and explanations to thought experiments that, as Elgin has also noted, play such a crucial role in the progress of science and which may also be described as 'beautiful' or 'elegant'. Thus she takes Galileo's 'falling bodies' thought experiment that spelled the end for the Aristotelian view of motion and which has been described as 'the most beautiful thought experiment ever devised' and asks what it is, precisely, that makes this example so beautiful. One option is to appeal to certain non-aesthetic features in order to explain our use of aesthetic terms in such cases. In the case of a concrete experiment, we may refer to its optimal use of minimal material, for example, and we can extend this to thought experiments, where the relevant economy can be cashed out in terms of the particulars that we are prescribed to imagine. Here Murphy draws a useful contrast with thought experiments that are described in negative terms, as in the example of Szilard's version of Maxwell's Demon, described as 'the worst thought experiment ever'. This exhibits an apparent misuse of certain idealisations, leading to confusion, or, in the case of Darwin's whale analogy, 'needlessly strange' explanations. However, she goes on to insist that we can understand the aesthetic evaluation of thought experiments in a broader sense that illustrates their commonalities with works of literature. This then reveals their differences from both theories and concrete experiments.

As Murphy notes, works of fiction have themselves been described as extended thought experiments that may reveal fundamental insights about both the world and ourselves. Greene's *The Third Man*, for example, has been portrayed as a thought experiment on the tension that may arise between maintaining loyalty to a friend and loyalty to a cause. Scientific thought experiments may also be presented in a narrative form, such as Newton's famous bucket experiment, and Elgin has argued that we can identify a continuity between concrete experiments, thought experiments and literary works. All involve a certain degree of control of a scenario as well as the use of idealisation, with differences within each case regarding how far and what ways they diverge from reality.

Moving in the other direction, from literature to science, Murphy considers how the aesthetic choices scientists make in the design of thought experiments contribute to their function in terms of communicating, convincing, or explaining a theory or phenomena to a scientific or a public community. Thus part of their value in the scientific context has to do with the features they share with literary works. However, it may be objected that there are clear disanalogies between the two. So, for example, Norton maintains that thought experiments are merely disguised arguments and hence their aesthetic qualities are dispensable. Similarly Egan argues that the purpose of a thought experiment is exhausted by its contribution to an argument. On the contrary, Murphy argues, although we may reconstruct thought experiments as arguments, in doing so we effectively lose sight of certain features that are crucial to the practice they are associated with. As Gendler has noted, the demonstrative force of the thought experiment is much diminished if we reconstruct it in the manner that Norton and Egan suggest. In the case of Galileo's falling bodies example, we lose the way in which tacit knowledge of how bodies fall block certain Aristotelian objections.

According to Murphy, it is through the introduction of the particulars and familiar objects, such as Galileo's towers and balls, that we engage with the thought experiment, and therefore come to understand what it and the relevant theorising is all about. Given that thought experiments depend upon our imaginative capacities, these features must obviously be carefully chosen.

Frigg and Nguyen have also pointed out that literary works and thought experiments differ in that interpretation is a much more flexible affair when it comes to the former as compared with the latter. Here Murphy appeals to the distinction between the description of a work, whether artistic or scientific and an interpretation of it. If we take account of this, the difference between artworks and scientific models, say, begins to evaporate, with many of the constraints on scientific models say, that have been attributed to interpretation falling under description, with similar constraints applying to literary works. Relatedly, Murphy notes, thought experiments may have a greater range of interpretation than one might initially suspect and indeed, it is part of their scientific value that they may be revised and their impact contested.

Finally, Murphy suggests that in comparing thought experiments and works of fiction, a certain degree of care needs to be taken when it comes to choosing the relevant comparators. Short stories and speculative fiction might be better choices than longer, realist works. Perhaps the most apposite examples are fables and parables, where the point is also to persuade or explain and a simplified scenario is presented, involving everyday or familiar objects and here Murphy notes the similarities with Cartwright's comparison between models and fables. Hence Murphy concludes, the differences between thought experiments and artworks don't account for the use of the former in scientific practice, and part of their value in this context includes the qualities that they share with the latter.

Both Arcangeli and Dokic, and Murphy emphasise the transformative function of aesthetic features, whether to do with the sublime or beauty, and indicate how they contribute to revolutionary theory change. In his article, 'Epistemic Radicals and the Vice of Arrogance as a Counterfeit to the Virtue of Assured Epistemic Ambition', Matthew Kieran also considers the innovative and revolutionary aspects of scientific practice and examines the features possessed by the 'epistemic radicals' who effect such dramatic shifts. As he notes, such features are often associated with epistemic vices, such as arrogance and competitiveness. Consider the great mathematician John Nash, for example, perhaps most famously known for his work in game theory, but who also made fundamental contributions to the study of partial differential equations and who was also renowned for his arrogance and self-confidence. As Kieran notes, studies show that many creative scientists exhibit such vices. However, this then generates a tension, since epistemic vice is supposed to tend towards epistemic failure, and yet if it is the mark of the epistemic radical how can being such a radical be deemed to be an epistemic good?

Kieran dissolves the tension by arguing that arrogance, for example, can be seen as a 'counterfeit virtue', in the sense that it has an overlapping behavioural profile with assured epistemic ambition that does not fall prey to the epistemic error and misdirection that are associated with arrogance. Conceptualising arrogance as an epistemic vice standing in a counterfeit relation to the true epistemic virtue of assured ambition in this way then explains why such epistemic radicals can be either heroes or villains.

Thus he begins by noting how arrogance may be bound up with the characteristics of an epistemic radical, by helping to generate ambition and drive, for example. Nevertheless, it can also generate reckless ambition that is manifested in carelessness over methods, taking short cuts and epistemic licence more generally, leading to projects failing. Here Kieran cites the example of Lysenko and, more recently, Stapel who faked his data to prove what he 'knew' to be true already.

The counterpart to this vice is the virtue of assured epistemic ambition, which involves not just the internalisation by the scientist of certain aims as having significant value or being valuable for their own sake and the commitment to appropriately pursue such aims, but also an epistemically permissible high degree of self-trust in presuming she has a good enough chance of realising them. Thus the Curies, for example, seem to have been driven by the aim of discovering knowledge for its own sake. Hopkins, on the other hand, dedicated his life to wiping out infectious tropical diseases in order to alleviate widespread suffering. However, as Kieran emphasises, it is not enough to have such lofty aims—one must have the right kind of commitment to them and go about achieving them in an appropriate manner.

This does not mean that the epistemically ambitious should not take risks; on the contrary, such people typically adopt new approaches in pursuit of their aims and have a justifiable trust in themselves in doing so. This in turn does not preclude self-questioning—indeed, according to Kieran what we often see is a virtuous cycle of self-development involving an ever-increasing 'upskilling' that places the scientist in a better position to realise great ambitions. As a result she will be well-situated to become an epistemic radical without succumbing to the tendency of the arrogant to commit errors or adopt misguided approaches.

Having said that, both those who possess the virtue of assured epistemic ambition and those who are arrogant may end up regarding only a comparatively select few people as their epistemic peers. They may both tend to dismiss the views of others, preferring to investigate the matter at hand for themselves. However, Kieran argues, those who are assuredly ambitious typically possess strengths that their arrogant counterparts do not have and are not susceptible to the failings of the latter. Examples include close-mindedness and a presumption to an entitlement of success that the ambitious can avoid through appropriate self-reflection.

Given all of this, Kieran concludes, we should regard such arrogance as a vice that is a counterfeit virtue to the true virtue of assured epistemic ambition. As he goes on to suggest, this may have practical implications, including the perpetration of epistemic injustice as in cases where people from disadvantaged groups are construed as arrogant when they manifest the epistemic profile of assured epistemic ambition precisely because that profile is in tension with the profile that is stereotypically assumed for members of that group. An obvious example here would be women in mathematics and science (or, indeed, philosophy). Furthermore, although normatively we might want to encourage the possession of such a virtue it might be objected that we don't want scientists to be radicals all the time—as Kuhn famously pointed out, much of scientific work involves problem solving and what he called 'normal' science. One's response to this depends on how one regards virtues more generally. The 'global' virtue theorist may insist that such a virtue is partly constitutive of what it is to be a good scientist but that it doesn't need to be manifested all the time. Of course, there are numerous examples of scientists who do not have the ambition of the epistemic radical but still produce good scientific work. As Kieran notes, a 'situational' approach acknowledges that all we need for scientific progress is for some individuals or teams to possess this virtue and that the scientific community in general can present a mix of epistemic 'moderates' as well as radicals.

Like Murphy, French is concerned with the comparison between scientific theories and certain artworks, in his case specifically musical works. With regard to the ontological status of both it has been suggested they should be viewed as abstract entities sitting in some equally abstract space, such as Popper's infamous 'World Three'. The worry with such views is that they have difficulty accommodating the heuristic processes by which theories are discovered—at what point in such a process, for example, does a theory appear in this abstract space? Here he explores an alternative view, due to Collingwood, that takes both artworks—and not just musical works and novels, but also paintings—and also scientific theories to be 'imaginary things'. This obviously allows for the accommodation of creativity and the heuristic process by which both artworks and theories are brought into being. However, it equally obviously suffers from the problem of inter-subjective inaccessibility.

Collingwood's solution is radical: the audience of a piece of music does not actually hear the artist's creation, rather they reconstruct it in their own imaginations. This then raises the further problem of whether that reconstruction could be said to be the same as the artist's work and here the issue of what counts as the identity conditions for artworks looms large. French canvases Wollheim's type-token account as well as Zemach's relative identity approach and concludes that neither is up to the job. As an alternative he suggests that we give up on establishing such conditions and accept that what is in each of the audience members' minds is different from what is in the artist's, and each other's, but that there is sufficient commonality for critical engagement to occur.

Interestingly, Collingwood himself compares a musical performance to a scientific presentation and argues likewise that the scientist's thesis is reconstructed in the minds of the audience members. Here too issues of identity arise: was the theory of Special Relativity that Einstein had in mind in his annus mirabilus of 1905 the same theory as the one that Planck subsequently presented? And were either the same as Lorentz's theory or the 'version' presented by Minkowski? At the very least it remains unclear how we are to answer these questions or the further one as to whether the reconstruction that we engage in when we read Einstein's paper corresponds to the theory *he* had in mind.

Again, French suggests, we should abandon the search for such identity conditions on theories and the associated attempts to place them in some ontological pigeonhole. Instead we should follow Ridley who urges philosophers of music to shift their focus to the practices of performances. Consideration of the corresponding scientific 'performances', in the form of lectures and conference presentations, for example, has been generally absent in the philosophy of science but historians have long noted the performative aspect of science.

If we then accept that there are no theories for which identity conditions need to be provided, we can take such performances and scientific practices in general as providing the truth-makers for claims that are putatively 'about' such theories. Thus, French argues, the statement that 'Special Relativity is empirically adequate' is made true by the complex set of practices involved in the testing of certain claims, such as, famously, those involving length contraction and time dilation. This can then be extended to claims such as 'Special Relativity is an elegant theory', which is made true by the relevant practices—so, for example, if elegance is understood as cashed out in terms of some combination of parsimony and power, then the statement is made true by the relevant practices, involving, for example, the ease of deduction of certain statements from the axioms or fundamental claims of the theory, the way in which a wide variety of claims (both theoretical and empirical) can be obtained from these axioms and so on.

French concludes by suggesting that this shift to practices may offer a third way between those who dismiss such aesthetic qualities as merely subjective and those who accord them a degree of objectivity: they are objective not in the sense of corresponding to features of the world but in that of being embodied in the relevant practices that scientists engage in, and here we can see connections with Ivanova's work, for example.

Although this is a diverse array of contributions, we hope to have indicated a number of core themes: the relationship between aesthetic qualities, such as beauty and sublimity, and their contribution to scientific aims such as understanding and truth, the possession of such qualities by different features in scientific practice, from theories and explanations to thought experiments, and the attitude toward such qualities by scientists themselves. And more generally, these themes entwine around the central relationship between the philosophy of art and the philosophy of science, across which various devices and approaches can be carried, from one side to the other, to the benefit of both.

Bibliography

Boden, M. (2003), *The Creative Mind: Myths and Mechanisms* (2nd ed.). Abingdon, Oxon: Routledge.

Bueno, O., et al. (2017), *Thinking About Science, Reflecting on Art: Bringing Aesthetics and Philosophy of Science Together*. Abingdon, Oxon: Routledge.

Currie, A. (2019), 'Creativity, Conservativeness and the Social Epistemology of Science', *Studies in the History and Philosophy of Science* (online first).
Elgin, C. (1991), 'Understanding: Art and Science', in P. French, T. Uehling and H. Wettstein (eds.), *Philosophy in the Arts, Midwest Studies in Philosophy*. Notre Dame, Indonesia: University of Notre Dame Press.
Frigg, R. and Hunter, M., eds. (2010), *Beyond Mimesis and Convention: Representation in Art and Science*. Boston Studies in the Philosophy of Science, Vol. 262. Berlin, Germany: Springer.
Heesen, R. (2018), 'Why the Reward Structure of Science Makes Reproducibility Problems Inevitable', *The Journal of Philosophy*, 115: 661–674.
Ivanova, M. (2017a), 'Aesthetic Values in Science', *Philosophy Compass*, 12: e12433. https://doi.org/10.1111/phc3.12433
Ivanova, M. (2017b), 'Poincaré's Aesthetics of Science', *Synthese*, 194: 2581–2594.
Livingston, P. (2009), 'Poincaré's Delicate Sieve: On Creativity and Constraints in the Arts', in M. Krausz, D. Dutton and K. Bardsley (eds.), *The Idea of Creativity*. Leiden: Brill.
McAllister, J. (1996), *Beauty and Revolution in Science*. Ithaca, New York: Cornell University Press.
Reichenbach, H. (1938), *Experience and Prediction*. Chicago, Illinois: Chicago University Press.
Schindler, S. (2018), *Theoretical Virtues in Science: Uncovering Reality Through Theory*. Cambridge, UK: Cambridge University Press.
Stump, D. (2007), 'Pierre Duhem's Virtue Epistemology', *Studies in History and Philosophy of Science*, 38: 149–159.
Suppes, P. (1960), 'A Comparison of the Meaning and Uses of Models in Mathematics and the Empirical Sciences', *Synthèse*, 12: 287–301.
Van Fraassen, B. (2008), *Scientific Representation: Paradoxes of Perspective*. Oxford, UK: Oxford University Press.

2 Epistemic Gatekeepers

The Role of Aesthetic Factors in Science

Catherine Z. Elgin

1 The Problem of the Aesthetic

Scientific theories, models, experiments, and the like are often subject to aesthetic assessment. This is uncontroversial. What is controversial is whether such assessments have any bearing on their epistemic standing. Is there any epistemically good reason to prefer an elegant experiment to an inelegant one, a beautiful theory to an ugly one, a streamlined model to one that seems more like a Rube Goldberg machine? What are we focusing on when we make such assessments?

If we seek to understand the role of aesthetic factors in science, it would be nice to have a criterion for the aesthetic. We do not have one. The history of aesthetics is littered with failed attempts to devise such a criterion. For our purposes, however, something less may suffice. All we really need is to identify a few factors that are plausibly construed as aesthetic. Then we can try to determine what, if anything, they contribute. If we find a contribution, we can attempt to identify other factors that function analogously. We do not need an exhaustive list of aesthetic factors; nor do we need to determine what precisely makes those factors aesthetic.

Clive Bell (1913) maintained that the aesthetic response to works in the visual arts consists in the apprehension and appreciation of *significant form*. He was concerned with significant visible forms—colors, contours, configurations, and the like. I suggest that many aesthetic responses to works in the sciences consist in the apprehension and appreciation of scientifically significant forms in a logical space. That space need not be physical, and the apprehension need not be sensory. The aesthetic properties that concern us then are formal properties of scientific artifacts such as theories, models, methods, and experiments. A form, let us say, is scientifically significant to the extent that it illuminates something that bears on the scientific acceptability of the item that displays that form. Precisely which forms these are, of course, varies from one context of inquiry to the next (see McAllister 2007). Nevertheless, the extrapolation of the idea of significant form makes it plausible that features like symmetry, simplicity, systematicity, and their opposites are aesthetic features of scientific artifacts.

Bell's criterion is known to be inadequate. It fails to capture features of works in the visual arts that are undeniably aesthetic. No doubt, my extrapolation is equally inadequate, and for the same reason. Still, I am not suggesting that significant form is a criterion for being aesthetic. I am suggesting that it is an aesthetic characteristic of some scientific artifacts. Just as Bell's criterion enables us to identify some aesthetically important features of works in the visual arts, my extrapolation enables us to identify some aesthetically important features of works of science. My purpose is to consider what functions such factors perform. For that, I need a few plausible candidates. I do not need (and cannot provide) a full criterion. Luckily, a sketch will do.

2 Truth or Thereabouts

According to a familiar criterion for scientific acceptability, aesthetic factors turn out to be either irrelevant or merely instrumental. Science, it is held, has a single overriding epistemic goal. Realists maintain that the goal is truth: science is successful when it reveals the truth or something close to the truth. Instruments and methods are scientifically valuable just because and just to the extent that they are truth conducive. Norms, standards, criteria, and the like are acceptable just in case and just to the extent that they promote the discovery of truth. Constructive empiricists maintain that the goal is empirical adequacy: science is successful when it achieves empirical adequacy or something close to it. Instruments and methods are scientifically valuable just because and just to the extent that they are conducive of the development of empirically adequate accounts. Norms, standards, criteria, and the like are acceptable just in case and just to the extent that they promote the development of empirically adequate accounts. Either way, on this picture if aesthetic factors contribute to science, their value is instrumental. They are epistemically valuable because and to the extent that they are indicative of the goal being realized or because and to the extent that they promote its realization. To streamline discussion, I will speak as though the goal is truth; a parallel argument holds if the goal is empirical adequacy.

On this view, the justification for preferring a beautiful, economical theory over an ugly, gerrymandered one with ad hoc excrescences is that the beautiful theory is more likely to be true (or approximately true) than its aesthetically unattractive rival. An elegant experiment is preferable because it is more likely to reveal the truth than one that proceeds by case hacking. If this is right, aesthetic assessments might play a useful diagnostic role. As scientists investigate a promising theory in depth, they may come to discern hidden beauties in it. The theories whose hidden beauties they appreciate are the ones that they are increasingly confident are likely to be true, or nearly so. That likelihood is why they keep working with them. And that scientists who have studied an issue in depth think that a theory is likely to be at least approximately true is reason for the rest of us to agree. If this is so, there may be a correlation between aesthetic assessments of currently credited theories and truth-related assessments. Perhaps the fact that aesthetically sensitive scientists find a theory beautiful is a good reason to consider it prima facie acceptable. If, however, beauty and truth do not run in tandem, this account holds that aesthetic judgments in science are at best epistemically idle. At worst they threaten to mislead.

Regrettably, there is no reason to believe that a scientific representation's aesthetic features correlate with the probability of its being true. Standard aesthetic assessments do not, so far as we can tell, track truth. We appreciate harmony, symmetry, elegance, and simplicity, all of which can be found in works of art that make no pretense of being true. Many are not even truth-apt. Indeed, those aesthetes who advocate 'art for art's sake' would maintain that genuine aesthetic value is rightly indifferent to truth. Nor is there any a priori reason to expect the phenomena to arrange themselves so as to align with our aesthetic preferences. That scientists who have studied a theory in depth think it is true may be some reason for the rest of us to think it is true. That they think it is beautiful is not.

Many manifestly false theories are beautiful. Aristotelian biology, for example, treats species as fixed: each species has its own essence, its own proper function, and its own distinctive telos, which both determines the good for its members and explains their behavior. This seems much lovelier than contemporary Darwinian biology, which acknowledges a large role for chance in the evolution of species, indeterminacy at the boundaries between species, a non-trivial measure of adaptive opportunism. It has nothing to say about what, if anything, constitutes the distinctive good for the members of a species. Caloric theory, which takes heat to be a smoothly flowing fluid, seems at least as lovely as the kinetic theory of heat, which maintains that individual gas molecules careen randomly about. Moreover, scientists may continue to consider an account beautiful even after it has been decisively rejected. Feynman, for example, initially thought that Feynman diagrams could explain why empty space is weightless. He continued to consider his account beautiful even after he refuted it (Wilczek 2016). Scientists do not automatically lose their aesthetic appreciation for the theories they leave behind.

But if aesthetic judgments in science are epistemically idle, why do we go on making them? Perhaps scientists' aesthetic assessments supervene on, are derivative from, or even are mere expressions of truth-related judgments. Then, in calling a theory beautiful or an experiment elegant, a scientist would be simply using an aesthetic label to characterize something that she considered scientifically estimable for other, legitimate—that is, truth-related—reasons. This seems to strip the terms from their aesthetic role. The label gets exported, but its aesthetic function is left behind. 'Beautiful' seems to mean just 'good of its kind' (see McAllister

1996: 77–81). If this is so, then there is no role for genuinely aesthetic assessment in science.

Perhaps aesthetic judgments in science play a heuristic role. (I admit we are getting a bit desperate here.) They might function as fast and frugal ways to make preliminary assessments of theories, experiments, models, and the like. But like the heuristics that figure in psychology's System I thinking, they are shortcuts that not infrequently lead us astray. They may be valuable for the way they enable us (with our limited minds) to come to a conclusion. But their utility is primarily practical and the heuristics are buggy. When you need a quick answer to 'Is this likely to be true?' ask yourself, 'Do you find this beautiful?' This strikes me as a dreadful idea, at least until we can find some reason to think that there is a correlation—even a loose correlation—between assessments of beauty and the probability of being true.

3 An Alternative

The problem with all of these proposals comes, I suggest, from thinking that if aesthetic factors are genuinely of value to science, it is because they somehow promote or sustain the justified conviction that the items that display them are true or nearly true. I will argue that aesthetic factors are integral to good science. They are not mere instruments. Nor is their utility primarily practical. But there is no reason to think that they themselves are truth-conducive. Rather, they bear on the acceptability of a scientific artifact—a theory, model, experiment or whatever. They do so, I suggest, by functioning as gatekeepers on acceptability: they play a regulative role.

The idea that factors that are not truth (or empirical adequacy) conducive are nevertheless integral to good science may seem anathema. Science, after all, seeks to understand the world. But the understanding that science delivers is embedded in models that are not, and do not purport to be, accurate representations of the phenomena they bear on (see Elgin 2017; Cartwright 1983). They simplify, streamline, augment, and omit. They are not true. Some, such as Snell's law, are not even nearly true. To be epistemically acceptable, a model or theory must properly answer to the phenomena. But properly answering to the phenomena is not the same as being a true or accurate representation of the phenomena. Nor is it the same as being an empirically adequate one. Models that diverge from the truth also diverge from empirical adequacy. The Lotka Volterra model of predator/prey relations, for example, construes predators as insatiably voracious. They are not. Even the hungriest shark eventually eats its fill and stops eating. Scientists already distance themselves from the idea that acceptable models must be true or empirically adequate. So the claim that the contribution of aesthetic factors to science is not a matter of their being directly or indirectly truth (or empirical adequacy) conducive is not as alarming as it might first appear.

Elsewhere, I have argued that an understanding of a topic consists of a systematically linked body of information in reflective equilibrium that is grounded in fact, is responsive to evidence and enables non-trivial inference and argument about a range of phenomena. That body of information is a network of epistemic commitments that includes beliefs about matters of fact, acceptances of models, idealizations, and thought experiments that are known to diverge from truth, as well as of methods, norms, and standards. Roughly, the idea is that to be justified in believing that the Higgs boson exists (or, for that matter, that the Loch Ness monster exists) requires a commitment to the methods, norms, standards, and background assumptions that figure in establishing its existence. Since understanding is, in the first instance, understanding of a topic or range of phenomena—not understanding of an individual matter of fact—the various elements of a system of thought must be mutually supportive. If an aesthetic factor such as symmetry or elegance is integral to a network of scientific commitments in reflective equilibrium, and could not be eliminated from the system without threatening or undermining the system's reflective equilibrium, then if that system is, by current standards, at least as good as any available alternative, the factor is epistemically justified.

Even if beauty is not exclusively in the eye of the beholder, it is hard to characterize. Nor is it wholly a matter of significant form, however broadly construed. So I will focus on other aesthetic factors that figure in science, ones that seem to be largely if not entirely matters of form: symmetry, systematicity, simplicity, and elegance. I am not saying that these factors are aesthetic per se. The only one with any claim to that status is elegance. My point is that their function in scientific acceptance is aesthetic. A model's displaying symmetry, an experiment's being simple, a consideration's weaving seamlessly into a network of mutually supportive commitments are the sorts of things that make them epistemically attractive in science.

4 Symmetry

Symmetry is a matter of structural invariance. A symmetrical item—an object or a law—retains its structure under transformations. Thus, for example, a cubical block displays rotational symmetry in that it retains its shape when rotated. The members of collection that share a symmetry have the same structure and retain that structure under the same transformations. They constitute an equivalence class. With regard to the transformation in question, they are interchangeable (van Fraassen 1989: 243). Things are not symmetrical *tout court*. They are symmetrical in one respect or another. So in devising a theory or model, questions arise: What sorts of symmetries are we interested in? Under what sorts of transformations should the objects of interest retain their structure? These questions bear on the sort of theory or account we want to devise. The answers mark out our categories and shape our theorizing. The symmetries that figure in science are not like rocks; we do not just stumble over them as we go about our lives. They are products of decisions. We configure the domain, deciding what structures we want to preserve, and under what transformations we want to preserve those structures. Maybe, for example, we seek to preserve rotational symmetry, but have no interest in color invariance. Then the block would count as symmetrical even if its faces were different colors.

It is no accident that scientific models display specific symmetries. We design them to do so. In designing a model or partitioning a domain in such a way that certain symmetries are displayed, we tentatively commit ourselves to the view that those symmetries are scientifically significant forms. What is the basis of their significance? We have no reason to think that a symmetrical model is more likely to fit its target than an asymmetrical one or that it is likely to fit its target better than one that lacks the symmetry in question. A critical question is how symmetries and asymmetries affect epistemic decisions.

Buridan's ass found himself equidistant from two identical, equally nourishing and tasty bales of hay. With respect to choice worthiness, the two bales were symmetrical. Nothing favored one over the other. Symmetry paralyzed the poor beast. Rather than starve, let us hope, he would eventually choose arbitrarily—perhaps, favoring the one on the right, even though there was absolutely no reason to prefer it to the one on the left. If he did, we could not understand his decision. We could understand that he chose—indeed, understand why he had to choose. But we could not understand why he chose the pile on the right. There was no reason. This is unsatisfactory, particularly if we are in the business of studying ass psychology (which is, these days, a growth industry). We might give him a bye if this were a one-off choice. That much arbitrariness in the mental life of an ass we might be willing to tolerate. But if he found himself in the same situation on multiple occasions and he chose the pile on the right significantly more often than the pile on the left, we would be apt to insist that there must be a reason. There must be something about the hay on the right or the ass's psychology, we would be apt to think, that accounts for the difference. There *must* be a tie-breaker. Then we would embark on a quest for hidden variables.

Another case: A fair coin is rotationally symmetrical. In the flip of such a coin, nothing favors heads over tails. Suppose we observed a single coin flipped seventeen times in a row. Each time, the coin came up heads. We know that in an infinite sequence of tosses of a fair coin, there is bound to be an interval where the coin comes up heads seventeen times in a row. Nevertheless, we would almost surely suppress that knowledge and judge that the coin was biased. We expect a fair coin to come up heads about half the time in even a short run of tosses. Should it not, we again seek a hidden variable—something that biases the coin.

There is, of course, no guarantee that we will find the hidden variable we seek. Maybe there is none to be found. My point is that in cases like these we treat the perception of symmetry and the perception of asymmetry differently. If the bale of hay on the right were fresher, or nearer, or composed of different sorts of hay from the one on the left, we would probably consider that a symmetry-breaking difference and think we understood the ass's behavior. If the coin came up HTTH THTH TTHH TTTHT (which specific sequence is, after all, no more or less probable than a sequence of seventeen heads in a row), we would judge the coin fair. But the fact that the all heads sequence makes the coin *look* asymmetrical arouses our suspicions. We do not think we understand how it happened. Something, we believe, needs to be explained.

The conviction that a range of phenomena displays a certain sort of symmetry is an initially tenable commitment (see Elgin 1996). In constructing a system in reflective equilibrium, that conviction has a slight and defeasible claim on our epistemic allegiance. We need a reason to give it up. Such reasons are often easy to find. We may discover that preserving a commitment proves too costly. Maybe the only way to preserve the conviction that the coin was biased or that the bale on the right was more desirable than the one on the left is to invoke occult forces for which we can find no independent evidence. Being unwilling to do that, we conclude that the coin was in fact fair, and that the ass's several choices were mutually independent and arbitrary—that our suspicions about the cases were unwarranted. Our quest for hidden variables in cases like these is evidence of our commitment to symmetry. We may find that we have to rethink the situation and our conviction that symmetry was broken where we thought it was. Or we may have to weaken our commitment to symmetry and recognize a measure of pure chance in nature. But we abandon the commitment with reluctance. We prefer symmetry-preserving accounts. Symmetry, I suggest, is an aesthetically pleasing feature. Our preference for it affects our behavior in accepting or rejecting certain findings.

5 Systematicity

Writing on the philosophy of history, Morton White (1965: 222–225), distinguishes between a chronicle and a history. A chronicle is just a list of unconnected, established facts about an historical episode. It makes no mention of dependence relations between facts; nor does it attempt to impose any order on them. A history is organized. It links the various facts to one another, displays dependency relations, imposes an explanatory framework in terms of which it makes sense that things played out as they did. Pretty clearly, the epistemic value of a history of a given event is correlated with the number and perceived importance of the facts in the corresponding chronicle that it accounts for. To be sure, the history need not mention every fact mentioned in that chronicle. But, one way or another, the history should accommodate the important facts on the list. A history of the siege of Stalingrad that left it a mystery why so many people died would be unsatisfactory.

Something similar holds in science. A science seeks systematic, integrated, mutually supportive accounts of the phenomena it investigates. It eschews danglers. A series of independently established truths about a domain might qualify as a natural chronicle. But a scientific account would be unsatisfactory if it left mysterious how those truths hang together.

Unaccommodated, isolated, but seemingly significant truths are considered problematic. Towards the end of the nineteenth century, celestial mechanics entertained a variety of increasingly strained hypotheses to accommodate the apparent anomaly in the perihelion of Mercury. But even at their most desperate, scientists did not advocate accepting:

> All planets except Mercury have regular Newtonian orbits; Mercury is just different.

Although the contention is true, was justified and confirmed by the observational evidence, and is not a Gettier case, it is unacceptable without an explanation of exactly why Mercury is different.[1]

Unexplained danglers are unacceptable. A question is considered open until putative exceptions to a proposed answer have been explained, or explained away. Unless apparent exceptions can be shown to be irrelevant, they are in one way or another expected to be woven into the scientific account. To be sure, there are anomalies—seemingly relevant issues that currently defy explanation or incorporation into our best available accounts. But anomalies are construed as outstanding debts. A scientific account is unsatisfactory to the extent that it lacks the resources to pay its debts. Moreover, once a phenomenon has been woven into an acceptable account, scientists are typically satisfied. Nothing more needs to be explained. Why Mercury's orbit is irregular was an outstanding problem for Newtonian mechanics. Why Mercury's orbit is regular is not a question for general relativity. Indeed, the question hardly makes sense. After all, given the theory, what would you expect?

Enthusiasm for systematicity runs deep. We want our fabric of scientific commitments to be tightly woven. Physicists seek a Grand Unified Theory out of a dissatisfaction with having to admit multiple fundamental forces into their ontology. It would be aesthetically more pleasing if there were only one (see Weinberg 1992). We dislike case hacking as a way of establishing an hypothesis because, it seems, the fact that the hypothesis can be shown, one by one, to apply to each individual case does not to our satisfaction show why it applies to all of them. The significant form of an acceptable scientific account consists in its being an interwoven (preferably tightly interwoven) fabric of epistemically interdependent commitments that has few, if any danglers.

6 Simplicity

Simplicity is complicated. There are numerous dimensions along which a theory or model might be simple or complex. Occam's Razor is traditionally formulated as a principle of ontological simplicity: Do not multiply entities beyond necessity. But, as Sober points out, there is a multiplicity of razors, each seemingly worthy of adoption (2015). There is no obvious general reason to favor ontological simplicity over, for example, axiomatic simplicity, syntactic simplicity, or inferential simplicity. Ontological simplicity consists in being committed to a minimal number of primitive entities or a minimal number of primitive kinds of entities. Axiomatic simplicity consists in containing a minimal number of axioms, fundamental laws, or postulates. Syntactic simplicity concerns the number and complexity of the basic syntactical units that figure in laws or axioms. Inferential simplicity concerns length of the derivations from the fundamental laws to other commitments of the theory. No doubt this list could be extended and the distinctions sharpened. For our purposes, the point is that these are all reasonable, objective, attractive features of theories and models. Moreover, they seem to have little to do with truth or empirical adequacy. There is, on the face of it, no reason to think that a theory with a minimal number of primitives or one that admits of streamlined inferences is more likely to be true than one that has a greater number of primitives, or one that requires long, convoluted inferences. Nor is there any reason to think that the prospects of empirical adequacy are enhanced by either sort of simplicity.

It might seem then that our preference for simplicity is grounded in intelligibility or tractability. Simpler theories and models are easier to handle. This is a pragmatic asset. If so, at least one reason why a simple theory is to be preferred over a complex one is that we are less likely to make mistakes in thinking with it or acting on it. The preference then is an accommodation to our human cognitive limitations. But some simplicity comes at the cost of intelligibility. Classical propositional logic needs only one primitive—the Sheffer stroke. In terms of ontological simplicity, Sheffer-stroke logic wins hands down. But it is hard to do logic using only the Sheffer stroke. The proofs are long; many steps seem counterintuitive; mistakes are apt to be made. As a practical matter, it is preferable to formulate propositional logic using at least two, and typically three, primitives. Standard formulations of propositional logic are moreover, inferentially simpler than Scheffer-stroke logic. Proofs are shorter, more direct, and more intuitive. If intelligibility and tractability are our guides, then a somewhat complicated theory may be more attractive than the simplest theory we can devise.

In the quest for simplicity, we face trade-offs. We readily sacrifice a measure of one sort of simplicity in order to gain a different sort. Thales hypothesized that everything is water (see Aristotle, 983b27–33). Della

Rocca (forthcoming) thinks that reality is one and indivisible. Albert suggests that reality consists in a single physical object—the universal wave function, or alternatively, the single universal particle (1996). All three theories achieve an admirable measure of ontological simplicity. Albert and Della Rocca contend that *au fond* there is only one thing; Thales contends that there is only one sort of thing. All face the enormous burden of accounting for the ways the one seems to configure itself into wildebeests and armadillos, starfish and galaxies, molecules and mountains, and so forth. The laws they need to invoke are apt to be exceedingly complex to account for the manifest diversity of appearances at every level, given that there is only one thing, or sort of thing to appeal to.

Still, we are not willing to jettison simplicity altogether. We reject theoretical excrescences. An electric current is standardly described as a flow of electrons. What would be wrong with elaborating this claim and saying that the electrons are propelled by tiny, undetectable gremlins nudging them along with hockey sticks? The answer is obvious: The standard account is simpler. The introduction of gremlins with hockey sticks is completely superfluous. It contributes nothing to the tenability of the account that contains it. Because it contributes nothing, the evidence for the original thesis does nothing to support the gremlin addendum. Nor do we have any independent evidence. That being so, we not only have no reason to accept the gremlin hypothesis; we have excellent reason to reject it.

This is an easy case, for the hypothesis in question is idle. It adds nothing. Things get trickier when an added hypothesis adds something, but not enough. If, to explain the behavior of Buridan's ass, we had to introduce an additional psycho-magnetic ⊠-force that operated only when an organism was equidistant between two equally appetizing alternatives, we would resist, and decide on balance that we should rest satisfied with the view that the ass's choices were arbitrary. In the absence of independent evidence of the existence of such a force, we would deem the cost in added complexity too high for the payoff it promises.

Ptolemaic astronomy was committed to the view that celestial objects display a complicated pattern of circular motions involving epicycles, equants, and deferents. It was important that all the motions be circular. Later astronomers, such as Galileo, thought the pattern was unduly complicated. The price the theory had to pay to preserve the geocentric framework was too high. Still, the theory was as simple as it could be if it was to accommodate the phenomena within a geocentric system. What the Ptolemaic astronomers never did, and would have been utterly unjustified in doing, was add additional (perhaps undetectable) motions to increase the number of dimensions along which celestial objects moved in circles. Science favors a minimalist aesthetics. Rococo additions have no place.

In both the case of electron flow and the case of Ptolemaic astronomy, it might seem that concerns for simplicity are presumptively truth-conducive.

The complexifying hypotheses we rejected are ones that, on the evidence available, we have no reason to consider true. But sometimes we favor simplicity over truth. We devise models that prescind from truth in order to achieve simplicity. Boyle's law—$pV = nRT$—falsifies the behavior of gas molecules by ignoring the attractive forces between them. For certain purposes, this is reasonable. The attractive forces are too weak to make a significant difference to the phenomena we seek to accommodate. Moreover, a more realistic model, such as the virial equation, is apt to occlude patterns that the more idealized model exemplifies. Where attractive forces between gas molecules are negligible, it is not just reasonable, it is advisable to neglect them—to omit them from the representations in terms of which we understand the phenomena. This is something I have argued elsewhere (see Elgin 2017). Here the important point is that streamlined models often embody an understanding of the phenomena, precisely because they are simplified. They omit what is negligible, or in Strevens's terms not a difference-maker (2008), thereby enabling us to apprehend and appreciate the significance of what is non-negligible, the difference-makers.

Science's preference for simplicity is rather inchoate, with different sorts of simplicity trading off against one another, and different types of simplicity predominating in different contexts (see Sober 2015). Still, we reject rococo accounts. The question is: why? During his scientific realist phase, Hilary Putnam ventured the hypothesis that the laws of nature could be no more complicated than differential equations (Putnam 1975: 309). It might seem that a hard-nosed scientific realist, like Putnam at the time, should endorse such a claim only if God assured him that the challenge of figuring out the way the world is would not be too difficult for humans to meet, in much the same way that a teacher might assure her students that the exam will not be too hard for them if they study assiduously. We have no such assurance. So if we have to back Putnam's hypothesis with truth-conducive reasons, it is unwarranted. We should admit that we have no clue how complicated the laws of nature are likely to be. Alternatively, however, we might take Putnam's proposal to be an aesthetic constraint on acceptance. Perhaps scientific investigators have such a strong aversion to mathematical complexity that they balk when the equations get too complicated. Being convinced that the complicated equations couldn't be right, they consider the investigation that led to them to still be open, and seek a way to either replace them by or reduce them to something simpler. Again there is no guarantee that they will succeed. But such an aesthetic preference would explain their efforts to come up with simpler laws.

7 Elegance

Elegance in science is a combination of effectiveness and economy. Effectiveness is an instrumental matter. There is something we seek to achieve and when we are effective our efforts pay off. The Miller-Urey experiment was effective in that it sought to demonstrate and succeeded in demonstrating that amino acids emerge from reactions of chemicals—ammonia, methane, hydrogen, and water—believed to be plentiful on Earth in prebiotic times. The four chemicals in plausible proportions were sealed in a chamber that was subjected to occasional sparks, which mimicked the effect of lightning. Over several days, a series of chemical reactions occurred, eventually yielding thirteen amino acids (see Ball 2005). The elegance of the experiment lies in its simplicity, which engenders a sense of inevitability. Given the experimental design, it appears, nothing but the four chemicals could account for the production of the amino acids. The result was reached with no extraneous theoretical, computational, or material factors. The background assumptions were clear and uncontroversial. The design was straightforward. Indeed, the experiment is so simple it almost looks like a nerdy high school science project.

There is no obvious reason why an elegant experiment is more likely to yield a truth, or an important truth, than an inelegant one. Still, elegance is an epistemically advantageous property. An elegant experiment makes manifest what it achieves and how it achieves what it does. It exemplifies its scientific contribution. An inelegant experiment might disclose the same truth, but we would have a harder time recognizing that or appreciating how it did so. If the inelegant experiment is sufficiently complicated, it invites the worry that unappreciated confounding factors, rather than the hypothesis being tested, account for the result. The elegant result is more illuminating. Either it readily integrates its result into a currently accepted account, or it manifestly poses a challenge to that account.

8 Optional Stops

So far, my claims have been largely descriptive. Scientists seek symmetry; they favor simplicity; they strive for systematicity; they appreciate elegance. I've urged that symmetry, simplicity, systematicity, and elegance are aesthetic properties, but I haven't yet said much about how they contribute to scientific understanding. One contribution has already been hinted at. Deviations from the ideals of symmetry, simplicity, and systematicity demand explanation. They pose a problem that ought to be addressed. Conformity to the ideals needs no explanation. Indeed, any attempt to explain cases that conform to our desiderata is apt to look a bit weird. No one asks why things behave as expected, even when aesthetic factors figure in the expectations. There is then an imbalance in our demands for explanation.

Scientists evidently observe what Nozick calls the Optional Stop Rule (Nozick 1981: 2). Inquiry has no foreordained stopping point. If we reach a result we find implausible or unpalatable, it is always open to us to conclude that there must be something wrong with the investigation, the reasoning process, or the background assumptions that led to it. This is what I suggested scientists who follow Putnam's recommendation do when they arrive at a mathematically complicated scientific law. Indeed, a scientist's reason for exercising the Optional Stop rule may itself be aesthetic. Weinberg (1992) suggests that the quest for a Grand Unified Theory is at least partly motivated by dissatisfaction with the idea that there is a multiplicity of mutually irreducible fundamental physical laws. Rather than accept a result and move on, we can decide to investigate a matter further. This may lead us to refine our methods, question our presuppositions, or look for hidden variables.

There is of course no guarantee that further investigation will lead to a conclusion we like better. It may be that the original inquiry was impeccable and the result, although unpalatable, was correct. Convinced that each individual event has a cause, a physicist might invoke the Optional Stop Rule and insist that there must be a reason why a particular radioactive particle decayed at time *t* and another, seemingly identical one did not. Rather than recognize that radioactive decay is stochastic, he might insist that scientists should keep looking. Once a no-hidden-variable theorem has been proven, it might seem, we reach a natural stopping point. What more could we want? The difficulty is that the Optional Stop Rule still applies. Rather than accept the conclusion that there is no hidden variable, we can pursue the suspicion that there is something wrong with the theorem's proof. Such resistance is not ruled out, even if there is nothing more to find.

The point is not that we are always right in our assessments. It is rather that a plausible, palatable result is, in large part because of its plausibility and palatability, apt to be deemed acceptable. Then inquiry with respect to that question ends. A sufficiently implausible or unpalatable result is apt to spark further inquiry. How could it be? The result is treated as a challenge. More work needs to be done.

The Optional Stop Rule might itself seem untenable. It seems rather unfair, even prejudicial, to subject objectionable results to greater scrutiny than attractive ones. But actually, the rule and the treatment it prescribes are reasonable. The results that strike us as plausible and palatable typically are ones that are readily integrated into accounts in reflective equilibrium. And they strengthen the accounts they are integrated into.

A constellation of epistemic commitments is in reflective equilibrium when its components are reasonable in light of one another and the constellation as a whole is at least as reasonable as any available alternative in light of our antecedent commitments. In constructing a system of thought we begin with whatever antecedent commitments (beliefs, norms, methods, goals, etc.) we take to bear on the topic we seek to understand and the sort of understanding we seek to achieve. These are a motley crew. They are apt to be mutually inconsistent, and even where consistent, non-cotenable. They are likely to be gappy, failing to cover matters we think they should cover. They frequently contain errors, omissions, and other cognitive infelicities. They are not acceptable as they stand. But they encapsulate our current understanding of the phenomena, the ways to investigate them, the norms that bear on acceptability, and so on. So we start with them. I call the starting points initially tenable commitments (see Elgin 1996). We correct, excise, extend, and amend them to bring them into accord. Although none of our initially tenable commitments is completely unrevisable, commitments have different degrees of cognitive inertia. We are more reluctant to abandon some than others. Even this reluctance is defeasible. Should it turn out that the price of retaining a commitment is too great—that is, should it turn out that we can only retain it by revising or rejecting other commitments that we think are collectively more worthy of acceptance—we will abandon even a commitment with considerable cognitive inertia (see Elgin 1996).

Aesthetic considerations may be initially tenable. At the outset we might, like Quine, simply have a fondness for desert landscapes. No matter. If we think that theories should be simple, or that laws should disclose symmetries, we can begin by favoring accounts with those features. In that case, we build into our theorizing a bias against complexity and asymmetry. This may seem question-begging, but it is not. Or anyway it is no more question-begging than a bias in favor of comprehensiveness or evidential adequacy. In any case, the bias is readily overrideable should it prove too costly. On the other hand, at the outset, we may lack aesthetic biases, having no preference for elegance, simplicity, and the rest. Then our early attempts at adjudication will be indifferent to such aesthetic considerations. But should we find that simplicity, elegance and the like are characteristics of the accounts we ultimately endorse, and discover that the closest competitors that lack those characteristics are, on the whole, less tenable, these aesthetic characteristics acquire the status of initially tenable commitments that (at least weakly) constrain future theorizing. Considerations that display these characteristics will, ceteris paribus, have an easier time gaining admission than those that lack them.

Still, to say that we would like our theory to display a certain aesthetic profile does not assure that we will get what we want. Every component of the system is subject to review. We may find that, for example, cost of simplicity is too high. If we have to sacrifice, say, a considerable measure of comprehensiveness or evidential adequacy to satisfy our current criterion for simplicity, we have a prima facie incentive to revise the criterion. We may abandon it completely; or we may restrict its scope, concluding that e.g., the payoff for recognizing a multiplicity of distinct elements has advantages that ontological monism cannot match. On the other hand, scientific developments may strengthen and reinforce an aesthetic commitment. Arguably symmetry was a rather peripheral epistemic value in classical physics. In quantum mechanics, it has moved to center stage.[2]

So we are within our epistemic rights to initially prefer theories, models, and experiments that display particular aesthetic profiles. When we do so, aesthetic factors play a regulatory role. They make no claim to track truth or empirical adequacy. Rather, their role as gatekeepers is to shape our accounts, encapsulating our evolving ways of framing our understanding so that the truths and untruths, the empirically adequate and empirically inadequate considerations that we accept are in reflective equilibrium, and therefore worthy of our reflective endorsement.

Notes

1. As it turned out, of course, Mercury is not relevantly different. The Newtonian theory that construed it as different was wrong.
2. I am grateful to Steven French for this point.

Bibliography

Albert, D. (1996), 'Elementary Quantum Metaphysics', in J. T. Cushing et al. (eds.), *Bohmian Mechanics and Quantum Theory: An Appraisal*. Dordrecht: Kluwer, pp. 277–284.

Aristotle. (1941), *Metaphysics: The Basic Works of Aristotle*, ed. Richard McKeon. New York: Random House, pp. 689–934.

Ball, P. (2005), *Elegant Solutions*. Cambridge: Royal Society of Chemistry.

Bell, C. (1913), *Art*. New York: Frederick A. Stokes.

Cartwright, N. (1983), *How the Laws of Physics Lie*. Clarendon.

Della Rocca, M. (forthcoming), *The Parmenidian Ascent*. Oxford University Press.

Elgin, C. (1996), *Considered Judgment*. Princeton University Press.

Elgin, C. (2017), *True Enough*. MIT Press.

McAllister, J. (1996), *Beauty and Revolution in Science*. Cornell University Press.

McAllister, J. (2007), 'Model Selection and the Multiplicity of Patterns in Empirical Data', *Philosophy of Science*, 74: 884–894.

Putnam, H. (1975), 'On Properties', in *Mathematics, Matter and Method*. Cambridge University Press, pp. 305–322.

Sober, E. (2015), *Ockham's Razors*. Cambridge University Press.

Strevens, M. (2008), *Depth*. Harvard University Press.

Van Fraassen, B. (1989), *Laws and Symmetry*. Clarendon.

Weinberg, S. (1992), *Dreams of a Final Theory*. Pantheon.

White, M. (1965), *The Foundations of Historical Knowledge*. Harper & Row.

Wilczek, F. (2016), 'How Feynman Diagrams Almost Saved Space', *Quanta Magazine*, July 5.

3 Getting the Picture

Towards a New Account of Scientific Understanding

Letitia Meynell

1 Introduction

In recent years there has been increasing interest in scientific understanding as an epistemic success term that is distinct from scientific knowledge (see, for example, De Regt, Leonelli and Eigner 2009). Although this literature is diverse, three dominant strands can be found that have rather deeper roots in the philosophy of science: understanding as unification (Friedman 1974; Kitcher 1981); understanding through mechanistic thinking as in certain types of causal modelling (Salmon 1998; Woodward 2003); and a kind of contextualist pluralist approach to understanding (De Regt and Dieks 2005; De Regt 2009, 2014), which is in some ways similar to the account offered by Nelson Goodman (1968, 1978) and Catherine Elgin (1997, 2004).[1] Proponents of these views often see them as complimentary or at least not contradictory. However, they have not yet been brought neatly together in a single account. This is what I propose to do. At the heart of my approach is the thought that we should treat the characteristic content of understanding as pictorial, in contrast to the characteristic content of knowledge, which is propositional. By virtue of the distinctive ways in which they present their content, epistemically efficacious pictures exemplify, unify, show mechanical (and other) causal relations, allow for multiple readings and facilitate the contextualisation of their content, while still having determinate content. In other words, features of pictorial content facilitate cognitive and evaluative procedures that are *characteristic* of understanding and have already been identified as such in the literature. That understanding should be treated as something like "getting the picture" is supported, albeit weakly, by multiple remarks, more or less well-developed, linking understanding to picturing that can also be found in the literature on scientific understanding, as we shall see.[2]

Many of these authors tie scientific understanding to explanation—so much so that some have argued that all current accounts of scientific understanding could be reduced to explanation without any significant remainder (Khalifa 2012). My approach loosens this connection; explanations are but one kind of thing that can be understood and many things, for instance, meanings and observations, can be understood despite not being explanations. Perhaps explanations are typically given with the intent of producing or enhancing understanding and often they succeed, but they can also be *known* or, indeed, *accepted* without being *understood*. Thus one of the things that is distinctive about my analysis is that my account of understanding is not parasitic on explanation.

Nonetheless, one cannot appreciate current theories of scientific understanding in the philosophy of science and their limitations without seeing their origins in theories of explanation, especially Carl Hempel's influential covering law model (Hempel 1962; Hempel and Oppenheim 1948). So, this is where we shall start. Hempel is, however, more than a convenient foil. He also specifies the philosophical project as one of explication, which is useful for teasing out just what I do and do not mean when I identify the *characteristic content* of understanding as pictorial. From there, we will briefly survey the main current views, each of which can be incorporated into an account of understanding as "getting the picture". The definitive features of understanding identified in the fourth section appear again when we analyse the particular way in which pictures present their content in the fifth. There, I explain how treating the characteristic content of understanding as pictorial helps to explicate it as a distinctive epistemic success term. In the last section before the conclusion I bring the discussion a little closer to the real world by exploring one pictorial example—Feynman diagrams—and another non-pictorial example—the understanding required when subjects give their informed consent to participating in research. The hope is that having two such different examples will speak to the general applicability of the account of understanding offered and also clarify that while the cognitive processes of understanding are those that characterise grasping the content of a picture this does not entail that the subjective content of understanding is necessarily an actual mental picture.

2 Understanding, Explanation and Hempel's Logicism

The influence of Hempel's covering law account of explanation can hardly be overstated; it is woven throughout the fabric of late twentieth and twenty-first century philosophy of science. Although, for the present project, Hempel usefully specifies the philosophical task, he also makes two deeply problematic assumptions. First, he treats 'understanding' as a by-product of explanation, creating a link between explanation and understanding that subsequently, has rarely been interrogated. Second, because he treats explanations as arguments, which are propositional in form with logical success conditions (i.e., truth and validity) he implies that the characteristic content of understanding takes on propositional and logical form and success conditions. This not only makes it difficult to see how understanding could be an epistemic success term that is significantly different from knowledge, but, as we shall see in section 4, it is difficult to square with the unificationist and causal mechanical views of scientific understanding that have become popular in the literature.

While readers of this volume are doubtless familiar with the covering law account of explanation, it is nonetheless worth reviewing, albeit briefly. For Hempel, scientific explanations take the form of arguments. An event, regularity or law is explained by a more general law when it takes the position of a conclusion in an argument expressible in first order logic, the premises of which contain that general law. In this way, Galileo's and Kepler's laws are explained by Newton's laws of motion by virtue of being deducible from them.[3] Deductive nomological (DN) explanations are the prototypical form of explanation-argument. To be adequate they must meet four criteria: (i) the explanandum must be a deductive consequence of the explanans; (ii) the explanans must contain at least one general law that is required for the derivation; (iii) the explanans must have empiricial content and thus be at least in principle testable; and, (iv) the sentences of the explanans must be true (Hempel and Oppenheim 1948: 137). Those arguments that meet the first three criteria but fail the fourth are potential explanations. Deductive statistical (DS) explanation-arguments—deducing one statistical law from others—employ the same formula, while the challenges of probability inflict a somewhat degenerate version on inductive statistical (IS) explanations, which are given to explain singular events using statistical laws (Hempel 1962).

Now, Hempel knows that when scientists explain phenomena they do not actually carefully devise logically sound arguments. The covering law model of explanation is a *rational reconstruction* and as such it is meant to exhibit the underlying rationale and logical structure of explanations, presenting an ideal that implicitly provides standards for critical appraisal (1962: 15–16). Through rational reconstruction the *concept* of 'explanation' is explicated. Explications take vague pre-theoretical concepts and refine them into something more precise, fruitful and (ideally) simple, without changing the meaning so drastically that it no longer pertains to the original applications of the pre-explication concept. Through providing an explication of scientific explanation Hempel not only produced the foil to which his successors have responded, but he defined the philosophical project in which they (we) are engaged as one of conceptual clarification through idealised rational reconstruction.

Given Hempel's own recognition of the relation between understanding and explanation (Hempel and Oppenheim 1948: 146; Hempel 1962: 9, 11), it may seem surprising that there is not a satisfying explication of *understanding* as a distinctive epistemic success term that can be found in, or even teased out of, Hempel's work. However, with a little reflection one can see why, from a Hempelian perspective, there may seem to be nothing of epistemic significance left to explicate beyond explanation itself. The two central epistemological concerns for logical empiricists were confirmation and explanation (Andersen and Hepburn 2015: §3). Confirmation, in its hypothetico-deductive form, constitutes justification for scientific *knowledge* claims—especially laws and theories—and explanation helps us *understand* phenomena. Although covering law explanations and hypothetico-deductive confirmations have fundamentally different aims and have faced rather different challenges and objections,[4] crucially they share the same logical structure. Explanation in its DN or DS form employs good old *modus ponens* (and even IS explanations come as close as possible while still being inductive (Hempel 1962: 13 ff.)). Confirmation employs exactly the same content as explanation in the same logical form, albeit with the indignity of affirming the consequent or recourse to the falsificationist gambit of denying that any empirical laws or generalisations are actually known (Popper 1959/1968: 278). For Hempel, and this is the crucial point for us, the sentences of the explanans and explanandum and their logical relations are the same as those that do the work in confirmation. Thus scientific knowledge and scientific understanding do not require separate logical analyses because they are merely different attitudes toward the same content. Moreover, this content and its internal logical relations are objective matters, rendering scientific knowledge and understanding objective matters. If confirmation and explanation require a similar type of assessment of exactly the same content then it is difficult to see what work understanding as a distinctive epistemic success term associated with explanation could be doing.

3 On the Characteristic Content of Epistemic Terms

Treating the characteristic content of understanding as having an entirely different form from knowledge promises to offer an analysis of understanding that neither threatens to reduce to or to replace knowledge. To begin to get a handle on what I mean by "characteristic content", it is perhaps useful to apply the phrase to the preceding discussion. According to Hempel, the *characteristic content* of understanding is propositional and thus has a logical form, just like explanation. This is not to suggest that Hempel is committed to any views about the propositional character of the mental content of epistemic subjects who claim to understand something. Here, the comparison with treatments of knowledge as a kind of propositional attitude is helpful. Though some epistemologists take this quite literally and maintain that all knowledge is constituted by actual mental attitudes toward propositions, others allow that in some cases knowledge can have non-propositional content. Nonetheless, such thinkers may offer *analyses* of knowledge that *treat* the content as propositional or maintain that it is reducible to, and thus analysable as, propositions.[5] There is nothing objectionable about doing so as long as the type of content that is relevant to theories of knowledge has the structure and success conditions that are characteristic of propositions as expressed by declarative sentences. Sentences are related to each other by the syntactic and semantic features that we associate with logic and it is these kinds of relations that are relevant to assessments of knowledge claims and justificatory conditions. In this way, one may *treat* the standard, representative content—the characteristic content—of knowledge as propositional without maintaining that *all* knowledge in reality comprises only mental attitudes toward propositions.

Similarly, I propose to treat pictures as the standard, representative content—the characteristic content—of understanding. Just as propositional form lends itself to the cognitive procedures relevant to producing knowledge, pictorial form lends itself to the cognitive procedures relevant to producing understanding. Like declarative sentences, the content of pictures is external and intersubjectively available to any competent witness. As the external form of declarative sentences and their internal and external logical relations are supposed to roughly line up with the internal content of what we know, so I take the external content of pictures to roughly line up with the internal content of what we understand. This is the sense in which the characteristic content of understanding is pictorial.

While Hempel's explicatory approach to explanation is a rough model for my explication of understanding, there are important disanalogies. Where Hempel maintains that actual scientific explanations are elliptical arguments that we can better analyse and assess by treating as if they were deductively valid arguments, I do not maintain that the actual mental content of cases of understanding is in any sense *inherently* pictorial but only that the cognitive procedures associated with understanding are more characteristic of those applicable to pictures than propositions. The type of relations between component parts of the content upon which the analysis and assessment of understanding as an epistemic success term rest are those characteristic of pictures, rather than propositions. Thus, we can better analyse and assess cases of understanding by treating their content in pictorial terms. If this is right, it is unsurprising that pictures are often effective ways of promoting and expressing understanding and that treatments of understanding in scientific epistemology often use pictorial terms.

4 Current Accounts of Scientific Understanding

With a clear articulation of the philosophical project and some of its context in hand, we are now in a position to survey some current theories of scientific understanding with an eye to drawing them together into a single coherent account. Proponents of the causal-mechanical and unificationist views have readily acknowledged the complementarity of these views, albeit without really explaining it. The contextualists, De Regt and Dieks (2005), bring the other views together in a sense, but only by compartmentalising them in separate contexts. Goodman and Elgin's approach is broad enough to encompass the others, but it lacks sufficient rigour to either produce the kind of explication that reveals the underlying structure of understanding or suggest implicit standards for critical appraisal. Each of the theories surveyed here moves away from Hempel's account of explanation and highlights an aspect of understanding favourable to the pictorial view that I offer in section 5.

4.1 *Kitcher's Unificationism*[6]

In "Explanatory Unification", Kitcher explicitly places himself in the Hempelian tradition, endorsing what he calls Hempel's "unofficial view". Explanation, he maintains, creates "an objective kind of insight that is achieved by systemic unification by exhibiting the phenomena as manifestations of common, underlying structures and processes" (Hempel quoted in Kitcher 1981: 508). Like Hempel, Kitcher places argument at the centre of his account. Certain arguments explain and promote understanding through unifying phenomena, not by way of nested logically sound deductions, but through a rather looser arrangement of propositions. According to Kitcher, the set of scientific facts at any given time has a set of arguments that are acceptable as the explanations for those facts. The unification of science is achieved by selecting from that "explanatory store" those arguments that fit the smallest number of generic argument patterns possible. In this way we understand the maximum number of facts through the smallest possible number of concepts and assumptions. Despite Kitcher's efforts to create a rigorous and systematic analysis, the notion of an argument pattern remains vague—at best it is merely suggestive. This muddies key parts of the analysis, such as what it means for a set of arguments to instantiate the same argument pattern or how to specify simpler, more generic, higher level argument patterns that subsume lower level patterns.

While Kitcher's adoption of a less rigorous notion of argument moves substantially away from Hempel's logicism, it is in his own examples where the plausibility of his account is tested and implicit support for a more radical pictorial account of understanding can be found. The first example is telling because it is implausible, in the face of standard explanations of the particular case considered, which typically employs pictures. It occurs as an illustration of explanation as an activity wherein one cogniser produces a kind of mental state—understanding—in another (p. 509). Kitcher imagines "a mythical Galileo confronted by a mythical fusilier who wants to know why his gun attains maximum range when it is mounted on a flat plain, if the barrel is elevated at 45° to the horizontal" (p. 510). According to Kitcher, mythical Galileo "selects a kinematical argument which shows that, for fixed velocity, an ideal projectile attains maximum range when the angle of elevation is 45°" (p. 510).

Now, while I suppose a mythical character can choose pretty much any method to explain various phenomena, the perusal of an introductory physics text or two (e.g., Jones and Childers 1990: 80; Giancoli 1998: 64; Knight 2008: 162) suggests that some kind of visual device is par for the course when explaining why a projectile achieves the maximum range when the angle of elevation is 45°. A picture is not, strictly speaking, necessary, especially if the fusilier has a good understanding of the relevant mathematics. Nevertheless, the geometry of the situation lends itself to pictorial explanation. A picture's explanatory or unifying power cannot be based on its argumentative form as it simply doesn't have one.

The fact that pictures aren't arguments (at least not in a logician's sense) seems lost on Kitcher in another example that he offers, this time one that explicitly identifies an image: Darwin's diagram from the *Origin of Species* (see Figure 3.1), and some fictional narratives that precede it. These together supposedly "[exhibit] a pattern of argument, which, [Darwin] maintains, can be instantiated, in principle, by a complete and rigorous derivation of descriptions of the characteristics of any current

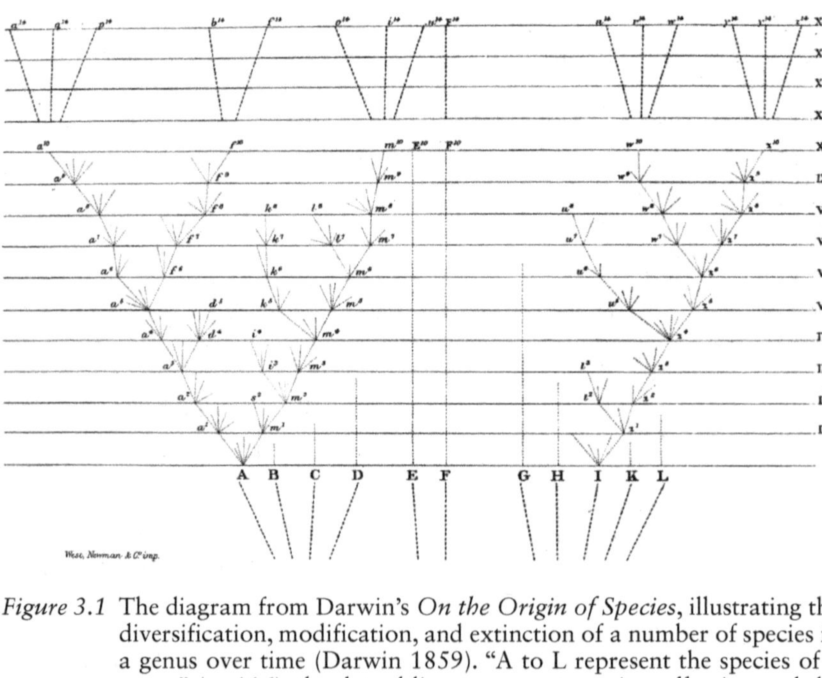

Figure 3.1 The diagram from Darwin's *On the Origin of Species*, illustrating the diversification, modification, and extinction of a number of species in a genus over time (Darwin 1859). "A to L represent the species of a genus" (p. 116), the dotted lines represent varying offspring and the horizontal lines 1000 (or perhaps 10,000) generations (p. 117)

Source: (Courtesy of Wellcome Library, London. Wellcome Images images@wellcome.ac.uk http://wellcomeimages.org Darwin's scheme showing the evolution of species. The Origin of Species Charles Robert Darwin Published: 1872 Copyrighted work available under Creative Commons Attribution only licence CC BY 4.0.).

species" (p. 515). While I would agree that this is an exemplary illustration of explanatory unification that enhances understanding, it is a real stretch to see either one of the narratives as an argument in Hempel's sense (though they contain arguments) and I am simply at a loss to guess what it means to say the diagram is an argument.

The first narrative considers populations of wolves with different prey evolving different morphologies that better suit the capture of that prey; the second considers the role of flower morphology and nectar production in the attraction of pollinating insects. Darwin's "imaginary illustrations" (1859: 90 ff.) describe how different selection pressures applied to subpopulations of a single variety can eventually result in a number of distinctive varieties. These little thought experiments, showing the inevitability of natural selection leading to distinct varieties, can be fitted into the diagram, but not as argument patterns. They are the kind of natural historical processes that might take one from, say, A to *a1* and *m1*. The diagram effectively invites the viewer to scale up these microevolutionary origins of varieties to origins of species, genera and so on, neatly adding extinction events and convergent evolution. The diagram does not present an argument pattern, but an abstract pattern of lineages that, though strictly speaking fictional, exemplify what Darwin supposed were the main kinds of important lineage relations, abstract morphological similarities, extinction events and temporal relations among taxa in the evolution of life.

While Kitcher sees himself as working in the Hempelian tradition, his move away from well-understood logical formulas to vaguely suggestive argument patterns is in fact a radical departure. In effect, he puts logical kinds into non-logical relations and it is difficult to know what to make of it. His examples take us a step further and suggest that he's not really talking about arguments at all. But there are some important lessons. After all, Darwin's figure does promote understanding through the unification of phenomena and representation of the common structures underlying diverse events. Moreover, this unification not only employs idealisations and generic explanatory structures but fits these idealised phenomena together into larger patterns. We will see these features again when we consider the way that pictures present their content in section 5.

4.2 *Woodward and Salmon's Causal Mechanical Account*

Although Wesley Salmon (1998) and James Woodward (2003) developed their causal-mechanical accounts separately, there are a number of similarities. Each sees himself as building on unificationist accounts, often uses pictorial language,[7] and emphasises action and experience (rather than propositions, symbols and arguments). Their focus on mechanism and causal processes brings to the fore the concrete and tangible, emphasising "how things work" and "what they're made of" (Salmon 1998: 88). Thus, Salmon maintains that "[scientific understanding] is the kind of understanding we achieve when we take apart an old-fashioned watch, with springs and cogged wheels, and successfully put it back together again, seeing how each part functions in relation to all the others" (p. 88). What we find in our experience, make use of in our actions and need to understand scientific phenomena are causal connections.

Woodward's causal account of explanation is fundamentally interactionist, stressing both the interactive capacities between different aspects of a phenomenon and scientists' interactions with their experimental apparatuses and objects. As with Kitcher, Woodward has a telling example involving a picture. To illustrate key concepts for his account—"invariant physical relations" (a rather less metaphysically fraught stand in for laws) and "intervention"—he introduces a familiar textbook diagram of a block on an inclined plane (see Figure 3.2). Woodward identifies the formulae for the frictional force, the gravitational force, the net force and the acceleration of the block, showing how each of these contains variables that relate to the others.

He writes:

> The account I defend takes this and other explanations to provide understanding by exhibiting a pattern of counterfactual dependence between explanans and explanandum—a pattern of counterfactual dependence of the special sort associated with relationships that are potentially exploitable for purposes of manipulation and control. In particular, the above explanation allows us to see how changing the values of various variables in [the equations for the frictional force, the gravitational force, the net force and the acceleration of the block] would result in changes in what it is we are trying to explain: the acceleration of the block. . . . [T]his what-if-things-had-been-different information is intimately connected to our judgements of causal and explanatory relevance.
>
> (2003: 13)

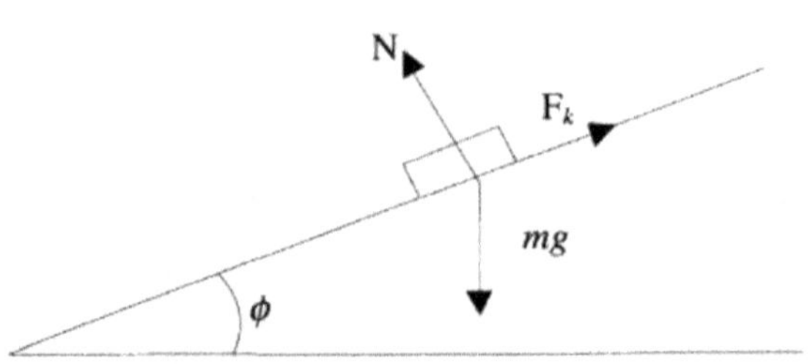

Figure 3.2 A block sliding down an inclined plane (from J. Woodward, *Making Things Happen: A Theory of Causal Explanation*, p. 13, Figure 1.4.1).

Copyright 2003 by Oxford University Press, Inc. Reproduced with permission of the Licensor through PLSclear.

Woodward emphasises that we can easily imagine how changing the value of one of the variables—say the angle of incline or the coefficient of kinetic friction—will change a number of the others. Indeed, we can make our what-if-things-had-been-different considerations pretty concrete, imagining replacing a glass surface with carpet or doing our experiments on the moon or in water, or even in water on the moon. The invariance of the relations means that interventions (be they imaginary or real) have predictable results, allowing us to manipulate and control phenomena.

Salmon, though originally attracted to a counterfactual theory, later turned to a looser account of causation (suggested by Dowe [1992]) that is friendlier to indeterministic causes (Salmon 1998: 133 ff.). Rejecting a Humean event-focused conceptualisation of cause and effect, Salmon instead treats causal processes in terms of conserved quantities (e.g., mass and charge), with interactions being manifested by an exchange of those conserved quantities (Salmon 1998: 253–254, following Dowe 1992: 210). Although the details are different, the basic causal-mechanical focus of the account remains much the same as Woodward's.

While I am sympathetic to this basic approach, there is an important proviso, concerning the relation between explanation and understanding. Explanations are focused on explananda—the point of the content of the explanation is to account for the explanadum. Understanding is a more flexible term. Where I might understand the acceleration of the block through Woodward's image, I might also use the same figure to understand the relation between the frictional force and the net force. Indeed, playing around with the various relations helps to reveal the limitations of the image. For instance, if the block only starts to move from the point pictured, then the static frictional force is also going to be relevant to its initial motion. Moreover, what is easy to miss when we look at this figure in the context of a philosophical example or even a physics textbook, is the way that a given cogniser orders the pictured content. What, in effect, they seek to explain will depend on what in the actual world they seek to understand or manipulate and the extent to which the pictured relations successfully translate into these applications. This freedom of interpretation is yet more obvious in Darwin's diagram and, indeed, Darwin demonstrates this by taking a number of passes through the phenomena represented in these lineages, each pass focusing on different aspects of the image, like differentiation of subpopulations over time, extinction, the different extent of variation, speciation within particular lineages and so forth. It is this type of imaginative freedom within the constraints of a given situation, thinking through the many counterfactual dependencies of a situation, that constitutes a deep understanding of that situation. This interpretive freedom is taken up by the contextualist approaches, to which we now turn.

4.3 Contextualism, Pluralism and Worldmaking

Though in some ways not a natural pairing, the recent work done by Henk De Regt and the account of understanding proposed by Nelson Goodman (1968, 1978) and developed by Catherine Elgin (1997, 2004, 2006) are brought together by their pragmatist, pluralist contextualism. These approaches differ from those discussed above, in that they do not treat understanding as a secondary by-product of an account of explanation. Though often neglected in more recent discussions, Goodman and Elgin were some of the first post-covering law philosophers to give a central place to understanding in scientific epistemology. Moreover, they are of particular interest in the current project as they tie understanding as an epistemic success term to both science and the representational arts. For Goodman and Elgin, both the sciences and the arts are ways of worldmaking, which employ various symbol systems (including natural language) to represent aspects of reality that produce understanding through exemplifying certain characteristics and relations. "Worlds" here should not be taken too literally. Goodman and Elgin are not defending a kind of idealism, though they do roundly reject any kind of monistic physicalism (Goodman 1978: 95). This is a many-things-go pluralism, not an anything goes relativism.

In *Ways of Worldmaking*, Goodman introduces the idea of "worlds" through a comparison with the notion of frames of reference, familiar from relativistic physics. Just as there are multiple frames of reference within which events can be described and no sensible, non-contextual or non-conventional answer to which is the "right" frame, so there are multiple "versions and visions" (1978: 2) of any given state of affairs, each of which may be right within its own terms. These versions reflect larger systems of belief or conceptual schemas, each with their own histories, applications and internal logics. We can only experience reality from where we stand; thus by necessity we start our artistic, scientific and philosophical work within the terms of some conceptual schema or other. By this application of concepts throughout perceptions, narratives and webs of belief we make our worlds. The manifest variety of irreducible and (more or less) mutually translatable descriptions of any given state of affairs leaves us with the epistemological work of determining the character and devising the norms of these different worlds.

Goodman offers a list of the processes of worldmaking, albeit an incomplete one. The processes are interconnected and overlapping but offer a kind of account of what it is to *understand* reality through making worlds. The first consists in "taking apart and putting together: on the one hand, of dividing wholes into parts and partitioning kinds into subspecies, analyzing complexes into component features, drawing distinctions; on the other hand, of composing wholes and kinds out of parts and members and subclasses, combining features into complexes and making connections" (1978: 7). Such dividing and composing has obvious implications for recognising permanence and change and, in turn, has ramifications for induction and prediction. Ordering, both logical and practical, also plays a crucial role in worldmaking. This activity goes from the highly conceptual ordering of colours, as on a spectrum, to the cognitive processes of ordering "a unified and comprehensive image of an object or a city from temporally and spatially and qualitatively heterogeneous observations and other items of information" (p. 13).

Deletion and supplementation are also crucial processes of worldmaking. Not only do we "find what we are prepared to find" (p. 14), but we also "dismiss as illusory or negligible what cannot be fitted into the architecture of the world we are building" (p. 14). This is nowhere more obvious than in the sciences, where theoretical assumptions and experimental design determine what counts as data, let alone "good" data, and what gets dismissed as outliers or anomalies or is simply invisible. In much the same vein, what counts as distortion and what counts as correction both depend on and reify a given world. Even for worlds that make the same corrections/deletions, the weighting of their content can render worlds radically different. Some kinds that are relevant in one world are present but deemed irrelevant in another. Goodman notes "Just as to stress all syllables is to stress none, so to take all classes as relevant kinds is to take none as such" (p. 11). Such weightings imply value hierarchies of "relevance, importance and utility" (p. 12).

It is in expressing these relevance hierarchies that *exemplification* is seen to be particularly important. While the details of exemplification are somewhat opaque (Vermeulen, Brun and Baumberger 2009), the basic idea is clear. Within a given symbol system certain symbols do not merely denote their referents but also instantiate and thus exemplify them. The famous example here is of a fabric swatch that exemplifies the relevant features of the bolt of fabric that it represents. Such exemplifications draw attention to themselves, expressing their own significance, and facilitate particular types of cognitive play and insight. As Elgin explains, "An exemplar is not a mere instance. It is a telling instance . . . by manipulating, scaffolding, or framing the context, the exemplar draws attention to particular aspects of itself" (2000: 221–222). Exemplification configures a world, unifying divergent phenomena as variations of a regular pattern and revealing interconnections that direct inspection of reality could not reveal.

The image of understanding that falls out of this account is of an active, on-going process of interpretation. There is no final achievement as there is with knowledge; it is a degreed notion with new understandings being "a springboard for further inquiry" (Elgin 1997: 79). The idea is not of a veil of perceptions sitting between the subject and the world, but a variety of veils of improvable conceptions through which we perceive the world. We start in a shared reality but because we bring different conceptual schemas to our navigation of reality we live in different worlds. Of course, there is considerable overlap between these worlds because they are ways of organising our shared reality and humans, at least, tend to be fairly similar in their cognitive and perceptual capacities. Moreover, many of us can move between worlds (albeit sometimes only with great effort); indeed, some of us have little option as we find ourselves positioned in our complex social lives in multiple incommensurable schemas of interpretation (Lugones 1987).

A similar type of pragmatist contextualism has recently been taken up by Henk De Regt with various coauthors (De Regt and Dieks 2005; De Regt et al. 2009; De Regt 2014). They maintain that there is nothing to guide the epistemic values chosen to mark understanding other than the interests and norms of the historical and disciplinary context. Key to their contextualism is the claim that there are no specific success standards characteristic of the achievement of understanding that are common across disciplines and across time (De Regt and Dieks 2005: 140). Nonetheless, although epistemic standards vary synchronically and diachronically, they are in some sense objective, in that they are not determined by the individual subject but are contextually given by the community.

While De Regt allows that understanding can be achieved through unification or a kind of causal thinking, he also thinks that "visualization is a very effective [though not indispensable] way to achieve scientific understanding" (2014: 378). He identifies the shared character with visual experiences as central to the efficacy of visualisation, noting, "As seeing is for humans arguably the most important way of grasping the world around us, it is not surprising that when we want to extend our grasp beyond what we directly observe, we prefer to rely on our well-developed visual skills and employ visualisation as a tool for understanding" (De Regt 2014: 394).[8] Although this seems like a promising start there is little real pay off. Rather than an analysis of visualisation, De Regt offers us a history lesson on the role of visualisability as a standard for intelligibility in twentieth-century particle physics. Though fascinating in its own right, this does little to clarify the nature of scientific understanding.

Though Elgin and Goodman give a central place to understanding in scientific epistemology, Elgin makes it clear that there is still a place, albeit secondary, for knowledge. Knowledge can be broken up into individual bits of information that are true, but as Elgin points out, it is rarely these individual facts that are of interest in science. Scientists investigate the relationships between these facts—how events happen and how systems and processes work. Indeed, often science isn't even particularly interested in truth, as when idealisations and abstractions, what Elgin calls "felicitous falsehoods," are deemed "true enough" and preferred over facts (2004). In essence, the processes of worldmaking are processes of sense-making—understanding at its core—and while individual facts (or claims that fit reality closely enough) have a role in this, it is how we put them together, what we identify as particularly meaningful, and how we project this into the future that really matters.

4.4 In Sum

Although each of the different approaches to understanding surveyed in this section has its weakness, I think they are all, at least in part, importantly right. Even if we don't accept Kitcher's *method* of unification through nested sentence patterns, the idea that unification by way of shared features and repeated patterns has an important role in understanding seems right. Goodman and Elgin's insight that we gain understanding not through always attending to every detail but drawing attention to sets of relevant features not only seems right, but resonates with Kitcher's view. But understanding, especially in certain scientific contexts, also requires an appreciation for the causal connections that determine past, future and counterfactual possibilities, as Salmon and Woodward note. While we might want to redirect their attention away from the provision of specific explanations, they are surely right that grasping complex causal relations is an important form of epistemic success. De Regt and Dieks are also right that different standards for what counts as understanding vary between times, disciplines and individuals, even if unification and an appreciation of causal relations are typically the values at play.

But if each of these accounts has something right, then the question is how can we bring them together? In the next section, I will show that if we treat the characteristic content of understanding as pictorial (a suggestion that clearly resonates with De Regt's emphasis on visualisation), we can see not only how the norms and cognitive processes of unification and causal-mechanical thinking are supported but also why a pluralist contextualist approach may be necessary.

5 Getting the Picture—Unifying Theories of Understanding

So, what is it about the way that pictures present their content that allows them to exemplify the cognitive processes and values characteristic of understanding? Happily, a good deal of work, identifying the distinctive character of pictures as content bearers has already been done (e.g., Meynell 2008, 2013, 2015; Pauwels 2006, Perini 2005, 2013; Tufte 2001; Walton 1990; Willats 1997). Two key features will hold our interest as they effectively unite the theories of scientific understanding discussed above—(i) their two-dimensional character and the related affinity with visual experiences and (ii) their role as representations, which, following Walton (1990), I take to be props for imagining.

5.1 The Two-Dimensional Character of Images and Unification

Pictures are a subset of a larger class of two-dimensional visual *images* (see Perini 2005; Meynell 2013). The significant form of images is distinct from that of sentences by virtue of the fact that images are two-dimensional[9] and so present their content all at once, in contrast to sentences, which are linear (Perini 2005). For sentences, two things determine content: (i) the spatial characteristics of the marks on a surface (e.g., ink marks on paper) determine the identity of characters (i.e., words), which stand for concrete objects, among other things, and then (ii) the order of the characters determines their syntactical or logical relations (Meynell 2013). Because the form of written words (and most spoken words, with the exception of those that are onomatapoic) is totally arbitrary with respect to their content, the significant form of language is retained from writing to speaking; the linear order of words is the only aspect of the form of propositions that determines content. In contrast to the linear form and syntactical relations of language, the significant form of an image is two-dimensional.[10]

The two-dimensional form of images requires and supports some of the cognitive procedures that the authors discussed above associate with scientific understanding. The linear character of language orders the content for the reader, but the viewers of an image must order the content for themselves. Clearly, they are highly constrained by the content of the image, but there is flexibility within these constraints. So, for instance, one can look at Darwin's image (Figure 3.1) with an eye to extinction and count the raw number of extinctions from the beginning to the end, or consider how the rate of extinction changes over time, or view the image without an eye to extinction at all. There are many different legitimate ways of ordering the content and in this way there are a number of versions of the content of the image with no acontextual way of deciding the right one. Viewers are forced to work out how these many aspects fit into the whole for themselves. This not only rewards Goodman's "taking apart and putting together," but also invites the comparison of parts. These processes allow the viewer to unify the content and "see the big picture," while appreciating the details. Notice that the two-dimensional character of the image, not its pictorial character, grounds this. If we wish to develop norms for critical appraisal we could do worse than start with Edward Tufte's list of features of "Graphical Excellence" in the now classic *Visual Display of Quantitative Information*, which has telling echoes of Kitcher's account. Tufte writes, "Excellence in statistical graphics consists of complex ideas communicated with clarity, precision, and efficiency. Graphical displays should [among other things]: make large data sets coherent; encourage the eye to compare different pieces of data; reveal the data at several levels of detail, from broad overview to the fine structure" (2001: 13). Of course, these norms can easily be modified for images that don't display data; they are simply good norms for two-dimensional images that are intended to display information and help the viewer understand it. As a class of images, pictures present their content in two dimensions and this feature alone grounds the ways that pictures promote the unification of information. Moreover, we also get a way of teasing out the hunch that understanding is a degreed notion. A low degree of understanding is achieved when one can grasp some of the legitimate ways of reading the image, a high degree of understanding is when one can grasp all (or at least most) of them and see how they fit together.

5.2 Pictures and Props for Imagining and Causation

With many information graphics, what I have identified elsewhere as "visual languages" (2013, 2015),[11] the spatial characteristics of the image bear no essential spatial or otherwise visual similarity or projective relation to the referents. This contrasts with pictures where at least some of the visible spatial (and sometimes colour) characteristics of the image are shared with the referent and the content of the image is determined in large part by these shared visible features (see Meynell 2013, 2015 for a provisional account and Willats 1997 for an exhaustive account of these relations). As we have seen, visual images of all kinds have certain epistemically relevant features, but pictures have additional ones that visual languages lack. Over and above the necessarily active and interested role of the epistemic subject in organising content, pictures, because of their spatial resemblance to the states of affairs that they represent, have implicit content derived from our knowledge of reality and causality. This is where De Regt's remarks on the intuitive connection between visualisation and understanding and the obvious association of visualisation[12] with visual experience really hits the ground. However, it also echoes Woodward's and Salmon's causal mechanical accounts. Insofar as our visual experience of reality is fundamentally causal so our comprehension of pictures that represent these states of affairs through their spatial features will also be causal.

The causal character of experience has a long history in the so-called western tradition, with perhaps its most prominent expression coming in Kant's Second Analogy (1929/1965, A189/B232 ff.). Even if we allow the cogency of a Humean scepticism regarding the *justification* of causal *inferences,* the role of causation in making our experiences intelligible can hardly be denied. Our experience of a ship sailing down a river is seamlessly organised by our cognitive and perceptual capacities in concert with the world, so that we can extrapolate from one moment both to what will happen in the next and what has happened in the past, and we make similar implicit judgements when looking at a picture that depicts the same moment. There is some flexibility in how we fill in the details of a picture—more than there is in the experience itself because we see one moment rather than a period of time and we are constrained by plausibility rather than actuality—but this is still quite limited. For instance, you and I may both be looking at Turner's *The Fighting Temeraire*, but you imagine the ship moving steadily and I see it as barely moving, you anticipate it is about to catch fire from the tug, I don't. We do not however, suppose that it might have been a bowl of petunias in the moment preceding, nor that it will turn into a bemused sperm whale in the next.[13] The implicit plausibility space of the picture effectively supports thinking though what-if-things-had-been-different scenarios, with Woodward's image of a block on the inclined plane as a case in point. We can, in our imaginations, remove and include various features and 'see' the many interrelated effects this will have on the pictured state of affairs. Which scenarios we envision will depend on our interests, but if we move outside the plausibility space, allowing anything to follow from anything else, we no longer get any understanding from the picture. It is the appreciation of the counterfactual constraints, the multiple mutual relationships and the limited flexibility within the constraints that constitute comprehending the picture and are characteristic of understanding a phenomenon, an event or a theory.

While it might be objected that sentences can also represent causes, sentences are actually pretty poor at specifying all of the antecedent conditions that are required for a certain state of affairs to come about and especially bad at specifying the various different ways contingent antecedent conditions might interact with each other. Pictures do this better because they show their content all at once in spatial relations that are relevant to causes. Of course, not all causes are easily visible, at least not in the right way or at the right scale, so pictures particularly lend themselves to representing mechanical relations. This, presumably, explains why pictures play such a crucial role in industrial design and architecture.

I have suggested that pictures have a kind of plausibility space that informs how we understand their content, but what determines the expectations and presuppositions that construct this space? Happily, Kendall Walton provides some tools for answering this question in *Mimesis as Make-Believe* (1990).[14] He introduces the idea of principles of generation (PGs), which are what a viewer brings to a work that allows them to glean the correct content. As he explains, "A principle is in force in a particular context if it is understood in that context that, given such-and-such circumstances, so and so is to be imagined . . . [P]rinciples of generation . . . constitute conditional prescriptions about what is to be imagined in what circumstances" (Walton 1990: 40–41). One will be misled if one takes the term "principle" too literally. Walton is referring to a cognitive miscellany that includes basic background beliefs (both explicit and tacit), acquired habits of mind (both those tacitly acquired and explicitly learned), conventions and explicit directions—many of the same kinds of processes that go into Goodman's worldmaking. Some of the background beliefs, habits of mind and conventions will be shared widely, general knowledge, as it were. Others will require discipline-specific expertise. So, for instance, it is common knowledge that DNA has a helical structure and most people can make sense of a simple graph with two axes, but knowing the C-value of the Takifugu genome and its significance or being able to make sense of a genetic ribbon diagram require expertise.

There is a final component of Walton's system that again resonates with, unifies and deepens the accounts given above. Walton distinguishes between the world of the work and the particular experience a viewer has of it. The work world is what should be imagined on viewing a picture given the PGs—i.e., the content as determined by PGs alone. Although this constrains what an actual viewer does imagine it does not fully determine it. Thus, despite the differences in how you and I might fill in the details of *The Fighting Temeraire* (discussed above), both are consistent with the world of the work though inconsistent with each other. This offers a way of thinking about the differences between subjective and objective (or at least intersubjective) scientific understanding (see De Regt and Dieks 2005). Though importantly distinct they are also intimately related. Even though the correct way to glean the content—the PGs—are determined at the level of the community it is the individual acts of applying these standards through which the community level norms are instantiated and reified. However, the very flexibility that individuals have within these norms provides the space for expansion and revision—possibilities of developing new ways of worldmaking—different understandings—that may better account for and exemplify the processes of interest.

6 Two Examples

The task of this chapter is to explicate scientific understanding as an epistemic success term, distinct from knowledge, by treating its *characteristic content* as pictorial as opposed to propositional. Were I to simply focus on pictorial examples in this section, I would risk creating the misleading impression that I think the mental content of understanding is necessarily imagistic or that all understanding requires some kind of picture. I do not. Certainly, it would count against my view if it made no sense of pictures at all. However, I remind the reader that it is the *cognitive procedures* that we associate with grasping the content of pictures that provide the key insights into the epistemic character of understanding.

What I propose to do in this section is to illustrate the account using two entirely different examples, one that *is* a famous type of scientific image—the Feynman diagram (FD)—and the other that has no pictorial content at all—the notion of understanding required to give informed consent to participate in research.[15] We will see that both exemplify the cognitive procedures that comprise scientific understanding as I have described it. While

I have elsewhere discussed in far greater detail the pictorial character of FDs (2008) as well as Feynman's interesting remarks about understanding and the ways in which FDs can elucidate interesting aspects of understanding (2018b),[16] constraints of space limit what follows to a few of those aspects of FDs that are directly relevant to the account above. In brief, my task here is to give a taste of how FDs help to unify phenomena in a cognitively accessible way, display complex causal connections within the phenomena, are produced by various PGs, and allow epistemic subjects to bring their own interests and commitments to envision quantum field theories in a variety of different ways. Similarly, the understanding required by a research participant in order to successfully give *informed* consent involves not only the acquisition but the integration of various different types of information in a way that appreciates how participation will and may cause various complications in their own life and what positive results might mean for people more generally. Having two such different examples speaks, I hope, to the general applicability of my account across scientific domains.

6.1 Feynman Diagrams

While there is still some controversy about whether FDs are pictorial at all (see, e.g., Brown 2018), here I will assume that they are indeed pictures of subatomic phenomena (a view I have defended elsewhere [2008]). This pictorial role is in addition to their uncontroversial role as calculational devices, devised to keep track of the long and cumbersome mathematical expressions in quantum electrodynamics. Feynman designed the diagrams so that each component corresponded to a specific mathematical expression. Briefly, the integral for calculating the probability of finding, say,

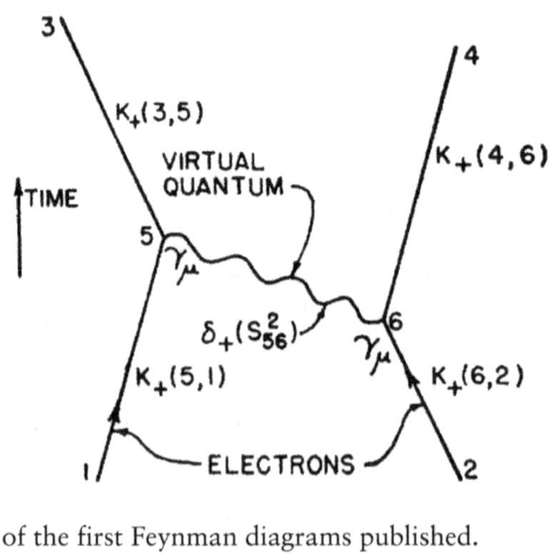

Figure 3.3 One of the first Feynman diagrams published.

Reprinted figure with permission from R.P. Feynman, Physical Review 76 (6), p. 772 (1949). Copyright 1949 by the American Physical Society.

two electrons at a particular place (3 and 4 on Figure 3.3), given their initial position (1 and 2 on Figure 3.3), was constructed by thinking of the photon mediating their interaction as taking all possible paths at once, with paths contributing less and less to the final value the less direct they were. A set of rules specified the correct construction of the diagrams (and, concomitantly, the integral). The most direct path (pictured in Figure 3.3) involves the emission and absorption of one photon. The next most direct paths involve the emission and absorption of two photons, but there are a number of different ways this can happen and so a number of different diagrams for the interaction involving two photons (Feynman 1949: 787), corresponding to a number of distinct expressions to be added to the integral. At the third order, three photons are involved and there is an even larger number of diagrams that can be drawn and yet more to be added to the integral, with the result becoming ever more precise (at least, that was the hope) and the integral becoming ever longer. Feynman diagrams simply made it easier to manage these massive calculations.

In addition to this, FDs are also pictures, providing a way of thinking about subatomic events, albeit in a manner that doesn't conform to classical physical intuitions. Some early quantum theorists thought that it was impossible to picture such events, but Feynman's own reflections make it clear that he disagreed. He was concerned not simply to solve mathematical problems but to understand "the physical picture," or, indeed, "pictures" that might be developed from the equations of quantum theory (Meynell 2018b: 464). In a letter to a friend, Feynman even waxed philosophical about the nature of understanding, identifying it with "seeing some of the qualitative consequences of the equations by some method other than solving them in detail" (Meynell 2018b: 465; Wüthrich 2011. Figure 4.1).

The picture of causal interactions between subatomic particles captured by the diagrams themselves—such as the interpretation of the positron as an electron going backward in time in the context of virtual electron-positron annihilation events (Meynell 2008: 51–52)—exemplifies the kind of counterfactual causal interactionism associated with Salmon's and Woodward's approach to understanding. This interpretation of the positron also illustrates just one way (among many) in which FDs served to unify phenomena by seamlessly bringing a theory of positrons into quantum electrodynamics, though, of course, the main achievement was to unify, in an intuitively graspable way, quantum theory with electromagnetism. That the diagrams have an intuitive appeal is sometimes thought to rest on their visual similarity to other images of subatomic particles, such as Minkowski diagrams and bubble chamber tracks (Meynell 2008: 45; Kaiser 2005: 186–187, 372–373). In Walton's terms, these are just some of the PGs—in this case, conventional and experimental images that were already meaningful to the trained expert eye in the 1940s—that physicists could draw on to make sense of FDs. Although the rules for construction were initially constrained by the mathematics, the representational potential of FDs opened up world-making opportunities, as theorists created stylistically similar depictions of subatomic systems where the mathematical method was clearly inapplicable (Kaiser 2005). Like Goodman and Elgin's exemplars, these images helped to configure the subatomic "world," suggesting various theoretical interconnections and patterns that might then be tested. Moreover, FDs facilitated the explanation of these events to non-experts, thus supporting superficial understandings of the theory to those without the mathematical knowledge required to enjoy an in-depth understanding.

Feynman's diagrammatic approach wasn't neatly deduced from theoretical precursors. Neither traditional theories of knowledge nor Hempelian explanation can fully capture its contribution to science. Rather, thinking about how FDs as pictures have been epistemically powerful lends itself to a distinctive account of epistemic success that, I think, can best be captured as understanding.

6.2 *Understanding Informed Consent*

The understanding required to give properly informed consent in research highlights rather different aspects of understanding, but nonetheless shares all of the features of *getting the picture* articulated above. As noted in *Canada's Tri-Council Policy Statement: Ethical Conduct for Research Involving Humans*,[17] "The key to informed consent is that prospective participants understand the information being conveyed to them by researchers" (2014: 30). This information must include known details like the purpose of the research, its duration, a description of research procedures and an explanation of the expectations of subjects and the details of their participation (29–31) ("including incentives for participants, reimbursement for participation-related expenses and compensation for injury" (29) and the circumstances that might lead to a subject's removal). Along with these knowns, researchers must disclose unknowns—the "reasonably foreseeable risks and potential benefits, both to the participants and in general" (30). Moreover, participants must be informed about the consent process itself, the possible commercialisation of the research and how the data collected might be used in the future, any potential conflicts of interest for the researchers, and who to contact "regarding possible ethical issues in the research" (29). Mere disclosure of this information or acceptance by the subject is not sufficient; it must be understood. Getting the picture usefully cashes out what it means to meet this condition.

First, and perhaps most obviously, the understanding at stake in informed consent admits of degrees. After all, a potential research subject might understand what they will be expected to do as participants but not grasp how these activities are connected to the risks of the research, thus not understanding the research well enough to give adequately informed consent. At the same time, it is widely recognised that complete understanding—say to the level of the researchers themselves—cannot reasonably be required for participants (Beauchamp and Childress 2009: 128) even if some subjects do in fact have this high level of understanding. Those participants whose profession means they have expert knowledge of aspects of the research process will likely find the information provided both easier to understand and have a deeper understanding of it than other participants. In Walton's terms, they already have a facility with some of the relevant principles of generation, like background beliefs and perhaps tacit expectations of how various research processes are worked out, that allow them to put together a picture of what the proposed research will (and could) mean to them as a participant. Nonetheless, ethical researchers will present the information in ordinary language and in such a way that any reasonable person can understand it. Indeed, the *Tri-Council Policy* recommends possibly supplementing consent forms with audio, visual or video aids to enhance understanding (30).

Interestingly, bioethicists have noted that too much information can actually undermine understanding (Beauchamp and Childress 2009: 130). Unlike knowledge, which grows with each new claim added, actually getting the picture requires the integration of information. Known facts and reasonably foreseeable possibilities and risks need to be brought together so that a subject can weigh them, see their various causal connections—how one thing might lead to another—and put themselves, their own lives and priorities, into the picture. Too many facts and possibilities, especially if they are not personally relevant to the particular subject can make it difficult to bring it all together to adequately understand what they might be getting themselves into.

Ultimately, what a research participant needs is to be able to unify all the information so that they can weigh the various pros and cons of the research but also grasp what the *qualitative consequences* of participating in the research will be in their own day-to-day lives for the duration of the study. They need to appreciate the various causal contingencies of the overall picture of what the research is (at least in terms of the procedures that involve them), including what the different end points might be so that by consenting they agree that they can live with these circumstances and these risks. This requires the ordering and juxtaposing of various expectations, commitments and values that constitute Elgin and Goodman's worldmaking as well as a kind of imaginative flexibility for thinking about the ways of fitting the rest of their lives around the research. Of course, this isn't the same as truly understanding the research itself but it will include at least a rough sketch of the science in question and a decent picture of what their participation in the research will mean to them.

7 Conclusion

When we consider treating the characteristic content of understanding as pictorial its distinctive epistemic character becomes clear. To understand something is to be able to see the component parts in relation to each other and as they relate to the whole. An epistemic subject who has mastered some material can take multiple paths through the content, though often their interests determine their focus. If the object of one's understanding is a state of affairs or phenomenon, as is typical in the sciences, causal relations are put into this kind of relational web. Those who deeply understand causal phenomena can see a kind of plausibility space where they can consider how things might have been different and appreciate, for any given change, what would follow from it with a view to the multiple relations that might be modified by the change. What determines this plausibility space are one's background beliefs (both tacit and explicit, ordinary and specialised); one's habits of mind (whether acquired with ease or through hard work); the applicable conventions; cognitive and perceptual capacities; and explicitly stipulated directions. Each one of these can be interrogated and thus the grounds of putative claims to understand can be assessed. From this perspective an explanation is a thread that we can disentangle from a much broader epistemic fabric of understanding. Through clearly articulating and closely examining this thread we get a better appreciation for the whole when we weave it back into the larger picture. Although there are still significant details to be worked out and many more insights from aesthetics and theories of pictorial content to be integrated, I hope to have shown that treating the characteristic content of understanding as pictorial offers a promising direction in developing a new account of scientific understanding.

Acknowledgements

I would like to thank Jonathan Longard for his research assistance as well as audiences in the Dalhousie Philosophy Department and at the *Bridging the Gap: Scientific Imagination Meets Aesthetic Imagination* Conference at the London School of Economics (especially Mike Stuart for inviting me) for their helpful feedback on earlier versions of this chapter. This research was supported by the Social Sciences and Humanities Research Council of Canada.

Notes

1. I borrow this characterisation of the literature from De Regt and Dieks (2005), though they do not themselves see the connection between their own view and that of Elgin and Goodman.
2. This chapter contextualises, develops, and defends some provisional remarks I make on the characteristic content of scientific understanding as pictorial in Meynell (2018b: 469 ff.).

3. Of course, this is not to endorse the idea that Galileo's or Kepler's laws are, strictly speaking, deducible from Newton's laws, which is a debate best left to those interested in realism and reductionism—topics beyond the scope of this chapter.
4. This can be seen in any philosophy of science textbook that has separate chapters or sections on confirmation and explanation (e.g., Salmon et al. 1992 or Curd, Cover and Pinnock 2012). Standard concerns about confirmation include underdetermination of theory by data, induction, and the paradox of the ravens, while standard concerns about explanation include worries about relevance and the distinction between explanation and prediction.
5. This kind of move is quite common among naturalised epistemologists who are interested in the idea of nonhuman knowledge. Often these theorists will present the belief content of a particular animal, say a rat, in terms of a belief that *p*, without wanting to suggest that the animal in question literally has an internal mental state where they consciously take an attitude to a sentence that expresses the proposition (see, for instance, Bermúdez 2003). Similarly, certain intellectualist approaches to knowing how might best be thought of as treating the content of "knowing how to φ" as propositional because some of the relevant success conditions and methods of analysis are those appropriate to propositional content (a number of the chapters in Bengson and Moffett 2011 seem to be navigating this terrain in various different ways, including the contribution by the editors themselves on "Nonpropositional intellectualism" (pp. 161–195)).
6. Friedman's analysis (1974) of the relation of understanding and explanation is mostly critical of others, with some fairly brief remarks about the key roles of unification and reduction for genuine scientific understanding. Kitcher's treatment is more exhaustive, so I follow it here.
7. For instance, Salmon, maintains that "we can say that we have *scientific understanding* of phenomena when we can fit them into the general scheme of things, that is, into the *scientific world-picture*" (1998: 88).
8. It is worth remembering the diversity of the human population in terms of both our cognitive and perceptual capacities, which suggests that different people will have various different capacities to form mental images and reason with them. Seeing is not "the most important way of grasping the world around us" for people with severe sight impairments and, indeed, some sighted people report being unable to visualise at all (Zeman, Dewar and Della Sala 2015).
9. The reader may worry that I seem to be neglecting two dimensions—a third spatial dimension (e.g., 3D models or sculptures) and time (e.g., film). However, the addition of these dimensions does not make a difference to the issue at hand. The relevant point here is the contrast with the linear (and so necessarily symbolic) character of language contra representational systems that present their content all at once. It may be objected that the temporal character of film is manifestly linear. However, although time is one dimensional this linear temporal component is added to the two-dimensional pictorial form which, frame-by-frame, presents its content all at once. The features arising from this pictorial character—the flexibility the viewer has to direct their attention to part-part, part-whole relations and so forth—remains. Hereon, I will assume for the sake of simplicity that what I say about two-dimensional images can also be extended to images of more dimensions.
10. This, of course, is something of an oversimplification. Pictographic languages and sign language have a variety of 'pictorial' content at the level of characters and sometimes phrases. Nonetheless, the syntax tends to be importantly linear; the ordering of the characters matters so that the meaning would

change if different readers started at various different places in a paragraph and ordered the content themselves.

11. It is worth noting that visual languages can have pictorial components (consider classic gender signs for washroom doors) but they serve the role of abstract symbol as they would have *exactly* the same referent were they replaced with a word (see Meynell 2015: 243–248 for a discussion of such cases). It is also worth noting that Perini (2005) and Pauwels (2006) carve up this landscape in ways that, in their gross features, are similar, though are rather different in the details.
12. Although the association is obvious, it is important not to overstate it. Sighted people differ substantially in the extent to which they find internal visualisation natural or even possible (Zeman et al. 2015). Nothing in my account, however, implies that those who lack developed internal visualisation capacities or are sight impaired will have a diminished capacity to understand.
13. Fans of science fiction will recognise the reference to the Infinite Improbability Drive from *The Hitchhiker's Guide to the Galaxy* (Adams 2005). The premise of this device is that it exploits quantum indeterminacy and its implication that many things that are, according to classical physics, absolutely impossible are only virtually impossible, and thus finitely improbable (pp. 89–90). The useful point here is that this kind of event, though strictly speaking possible, is nonetheless implausible, indicating that possibility and probability, which are reasonably well-theorised, are distinct from plausibility. How exactly they differ and how they are related is beyond the scope of the current discussion.
14. I have extended this theory to scientific images in various publications and I draw heavily from this work (2008, 2013, 2015, 2018b).
15. I thank Kirstin Borgerson for suggesting this example in conversation.
16. Meynell (2018b) can be read as an extended illustration of the account of understanding that I defend here.
17. The *Tri-Council Policy* is typical of national research policies on consent, roughly following and expanding on the relevant sections of the World Medical Association's *Declaration of Helsinki*.

Bibliography

Adams, D. (2005), *The Hitchhiker's Guide to the Galaxy*. Del Rey Books.

Andersen, H. and Hepburn, B. (2015), 'Scientific Method', in E. N. Zalta (ed.), *The Stanford Encyclopedia of Philosophy*, Summer 2016 ed. https://plato.stanford.edu/archives/sum2016/entries/scientific-method/.

Beauchamp, T. and Childress, J. (2009), *Principles of Biomedical Ethics*, 6th ed. Oxford University Press.

Bengson, B. and Moffett, M., eds. (2011), *Knowing How: Essays on Knowledge, Mind, and Action*. Oxford University Press.

Bermúdez, J. (2003), *Thinking Without Words*. Oxford University Press.

Brown, J. R. (2018), 'How Do Feynman Diagrams Work?', *Perspectives on Science*, 26: 423–442.

Canadian Institutes of Health Research, Natural Sciences and Engineering Research Council of Canada, and Social Sciences and Humanities Research Council of Canada. (2014), *Tri-Council Policy Statement: Ethical Conduct for Research Involving Humans*. Online at www.pre.ethics.gc.ca/pdf/eng/tcps2-2014/TCPS_2_FINAL_Web.pdf

Curd, M., Cover, J. A. and Pinnock, C. (2012), *Philosophy of Science: The Central Issues*, 2nd ed. W.W. Norton and Company.

Darwin, C. (1859), *The Origin of Species by Means of Natural Selection or the Preservation of Favoured Races in the Struggle for Life*. London: John Murray.

De Regt, H. (2014), 'Visualization as a Tool for Understanding', *Perspectives in Science*, 22: 377–396.

De Regt, H. and Dieks, D. (2005), 'A Contextual Approach to Scientific Understanding', *Synthese*, 144: 137–170.

De Regt, H., Leonelli, S. and Eigner, K. (2009), 'Focusing on Scientific Understanding', in H. De Regt, S. Leonelli and K. Eigner (eds.), *Scientific Understanding: Philosophical Perspectives*. University of Pittsburgh Press, pp. 1–17.

Dowe, P. (1992), 'Wesley Salmon's Process Theory of Causality and the Conserved Quantity Theory', *Philosophy of Science*, 59: 195–216.

Elgin, C. Z. (1997), 'Relocating Aesthetics: Goodman's Epistemic Turn', in C. Z. Elgin (ed.), *Between the Absolute and the Arbitrary*. Cornell University Press, pp. 63–80.

Elgin, C. Z. (2000), 'Reorienting Aesthetics, Reconceiving Cognition', *The Journal of Aesthetics and Art Criticism*, 58: 219–225.

Elgin, C. Z. (2004), 'True Enough', *Philosophical Issues*, 14: 113–121.

Elgin, C. Z. (2006), 'From Knowledge to Understanding', in S. Hetherington (ed.), *Epistemology Futures*. Clarendon Press, pp. 199–215.

Feynman, R. (1949), 'Space-Time Approach to Quantum Electrodynamics', *Physical Review*, 76: 769–789.

Friedman, M. (1974), 'Explanation and Scientific Understanding', *Journal of Philosophy*, 71: 5–19.

Giancoli, D. (1998), *Physics: Principles With Applications*, 5th ed. Prentice Hall.

Goodman, N. (1968), *Languages of Art: An Approach to a Theory of Symbols*. Bobbs-Merrill.

Goodman, N. (1978), *Ways of Worldmaking*. Hackett Publishing.

Hempel, C. (1962), 'Two Basic Types of Scientific Explanation', in R. G. Colodny (ed.), *Frontiers of Science and Philosophy*. Allen and Unwin and University of Pittsburgh Press, pp. 9–19.

Hempel, C. (1965), *Aspects of Scientific Explanation the Other Essays in the Philosophy of Science*. New York: The Free Press.

Hempel, C. and Oppenheim, P. (1948), 'Studies in the Logic of Explanation', *Philosophy of Science*, 15: 135–175.

Jones, E. and Childers, R. (1990), *Contemporary College Physics*. Addison-Wesley Publishing.

Kaiser, D. (2005), *Drawing Theories Apart: The Dispersion of Feynman Diagrams in Postwar Physics*. University of Chicago Press.

Kant, I. (1929/1965), *Critique of Pure Reason*, trans. N. Kemp-Smith. St. Martin's Press.

Khalifa, K. (2012), 'Inaugurating Understanding or Repackaging of Explanation?', *Philosophy of Science*, 79: 15–37.

Kitcher, P. (1981), 'Explanatory Unification', *Philosophy of Science*, 48: 507–531.

Knight, R. (2008), *Physics for Scientists and Engineers: A Strategic Approach with Modern Physics*. San Francisco: Pearson.

Lugones, M. (1987), 'Playfulness, 'World'-Travelling, and Loving Perception', *Hypatia*, 2: 3–19.

Meynell, L. (2008), 'Why Feynman Diagrams Represent', *International Studies in the Philosophy of Science*, 22: 39–59.

Meynell, L. (2013), 'Parsing Pictures: On Analyzing the Content of Images in Science', *Knowledge Engineering Review*, 28: 327–345.

Meynell, L. (2015), 'See What I Mean? On Developing Norms for the Production and Publication of Scientific Images', in J. Kanjirakkat, G. McOuat and S. Sarukkai (eds.), *Narratives of Science and Nature: East and West*. Delhi: Routledge, pp. 239–260.

Meynell, L. (2018a), 'Images and Imagination in Thought Experiments', in M. Stuart, Y. Fehige and J. R. Brown (eds.), *The Routledge Companion to Thought Experiments*. Routledge, pp. 498–511.

Meynell, L. (2018b), 'Picturing Feynman Diagrams and the Epistemology of Understanding', *Perspectives on Science*, 26: 459–481.

Pauwels, L. (2006), 'A Theoretical Framework for Assessing Visual Practices in Knowledge Building and Science Communication', in L. Pauwels (ed.), *Visual Cultures of Science Communication: Rethinking Representational Practices in Knowledge Building and Science Communication*. Dartmouth College Press, pp. 1–25.

Perini, L. (2005), 'The Truth in Pictures', *Philosophy of Science*, 72: 262–285.

Perini, L. (2013), 'Diagrams in Biology', *Knowledge Engineering Review*, 28: 273–286.

Popper, K. (1959/1968), *The Logic of Scientific Discovery*. Harper and Row.

Salmon, M., et al., eds. (1992), *Introduction to the Philosophy of Science*. Englewood Cliffs, NJ. Hackett Publishing Company.

Salmon, W. (1998), *Causality and Explanation*. Oxford University Press.

Tufte, E. (2001), *The Visual Display of Quantitative Data*, 2nd ed. Cheshire, CT: The Graphics Press.

Vermeulen, I., Brun, G. and Baumberger, C. (2009), 'Five Ways of (Not) Defining Exemplification', in G. Ernst, J. Steinbrenner and O. Scholz (eds.), *From Logic to Art: Themes From Nelson Goodman*. Ontos Verlag, pp. 219–250.

Walton, K. (1990), *Mimesis as Make-Believe: On the Foundations of the Representational Arts*. Harvard University Press.

Willats, J. (1997), *Art and Representation: New Principles in the Analysis of Pictures*. Princeton University Press.

Woodward, J. (2003), *Making Things Happen: A Theory of Causal Explanation*. Oxford University Press.

Wüthrich, A. (2011), *The Genesis of Feynman Diagrams*, Kindle ed. Springer.

Zeman, A., Dewar, M. and Della Sala, S. (2015), 'Lives Without Imagery—Congenital Aphantasia', *Cortex*, 73: 378–380.

4 Imagination, Aesthetic Feelings, and Scientific Reasoning

Cain Todd

1 Introduction

Nobody denies that thought experiments (TE) and scientific models (SM) play a crucial role in scientific reasoning. Further, it seems undeniable that TE and SM centrally involve the imagination. It is, however, in considering the nature of this involvement that a series of questions and problems arise. Many of these are epistemic: given the kind of activity imagining is, how could it possibly help to reveal empirical facts about the world? How could scientific beliefs formed via imagination possibly be justified? Some questions, however, are non-epistemic: what is a scientific model? What sorts of imaginings are involved in TE and SM? What exactly is the imagination? Clearly, answers to the epistemic questions depend in part on the answers to the non-epistemic questions. Yet these latter questions seem so difficult, and the various available answers so inconclusive, that the enterprise of understanding the nature of TE and SM, and hence the role of imagination in scientific reasoning can appear impossibly daunting.

In order to make these issues a little more tractable, the present chapter will focus primarily on just one form of imagining involved in TE and SM, the kind that is essentially imagistic and is often called 'visualisation'. The main questions to be considered, then, are: 1) what role does visualisation play in scientific reasoning? and hence 2) what epistemic credentials, if any, does visualising have in scientific reasoning? As we shall see, however, addressing these issues will involve saying quite a bit about how visualisation—which I will also refer to as 'sensory imagining'—relates to non-imagistic propositional imagining. In short, the connection between imagination and feeling (or 'affect') is central to understanding the cognitive value of SM.

Two further qualifications: I will only be concerned with theoretical and not physical models, and I will understand SM as being a type of TE. I do not deny that the success of some SM is independent of any visualisation that may take place. I will argue only that visualisation, and other types of imagery, can play a crucial epistemic role in SM. There may be good reasons to distinguish TE from SM, but I suspect that visualisation plays the same sort of epistemic role in each. For this reason, unless I refer to them explicitly, I will often talk of the role of visualisation in 'scientific reasoning' and I will take it that the conclusions reached can apply equally to SM, also to TE.[1]

Although some philosophers have denied that imagination plays an important role in scientific reasoning, many philosophers have come to accept that it does, and several have recently defended the epistemic value of imagining in SM. Nonetheless, it is fair to say that with few exceptions the role of imagery has been entirely downplayed—a mere psychological embellishment that itself does no epistemic work—or regarded with deep suspicion. Yet, surprisingly, philosophers working in this area generally have little or nothing to say about the nature of imagery. I do not think that the role of imagery in scientific reasoning can be so easily ignored, and my main goal will be to show how visualisation can play an important epistemic role here. Indeed, I shall argue that current accounts of SM that rely solely or primarily on non-imagistic imaginings fail to account adequately for their ineliminability and success in scientific reasoning. The way ahead will, though, be a little convoluted, because consideration of the relatively *non-transparent* nature of images poses a serious problem for the idea that visualising could possibly offer any epistemic insight into physical phenomena.

In brief, two key moves need to be made to secure the epistemic standing of sensory imagining in SM. First, in exploring the differences between the transparency of perceptual and imaginative experiences, we need a proper appraisal of the relationship between sensory imagining and affect, and in particular the affective feelings that are manifestations of aesthetic experience. Second, we need to focus our attention on the notion of understanding, rather than truth or knowledge.

2 Fiction and Scientific Models

The most prominent current views of SM that acknowledge the central role of imagination appeal to our engagement with fiction in order to illuminate this role. Toon (2016) helpfully divides such 'fictional' theories into Direct and Indirect. The latter are indirect in the sense that they require us to think of SM as like a work of fiction, or a fictional world, which we examine in order to find out what its properties are and which we then apply to the real world, of which it is a representation. The indirect view assumes, then, that the properties of SM go beyond those that are explicitly specified in the description of SM, and that these properties can be used to give insight into the real world in virtue, somehow, of representing it.

The most comprehensive attempt to explain how this is possible is given by Roman Frigg (2010a, 2010b), who appeals to Walton's well-known theory of fiction as make-believe. For Walton, fictional representations are *props* in games of make-believe; that is, fictional works (qua games of pretence) prescribe us to imagine certain things, in line with the propositions explicitly given in the work (the description of the game world), but also in accordance with certain *principles of generation* that constitute the rules of the game. Notoriously difficult to articulate precisely or fully, principles of generation will normally include, for example, and unless stated otherwise, a 'reality principle', such that for the situations and events not explicitly described in the work, we are to imagine that the world of the work operates more or less as the real world does. For instance, if Sherlock Holmes is described as travelling to Dorset, we are not supposed to imagine that he got there by hovercraft. We are supposed to imagine that the city he lives in is just as Victorian London would have been at the time, and so on. Thus, if a proposition is prescribed to be imagined in a game of make-believe, then it is fictional in the relevant game i.e. 'true in the fiction'. Frigg's idea then is that SM are props in games of make-believe.

When we read the model description for the Newtonian model of the solar system, Frigg states, "we imagine an entity which has all the properties that the description specifies. The result of this process is the *model-system*, the fictional scenario which is the vehicle of our reasoning: an imagined entity consisting of two spheres, etc" (2010b: 133). After imagining her model system, the scientist goes on to connect it to the real system. In this case, for example, she might specify that "the sphere with mass *me* in the model-system corresponds to the earth and the sphere with mass *ms* to the sun" (2010b: 134). Once this is done, she can "start translating facts about the model system into claims about the world" (2010b: 135).

The Direct Fiction view also draws on the Waltonian framework of make-believe—model descriptions are 'props' in a game of make-believe that prescribe imaginings about reality—but employs an analogy with fictions about real world events and characters, rather than fictions *tout court*. As Toon (2016) says of his own position, the main contrast with the Indirect view is that his does not posit any model systems representing the real world. Rather, scientists represent the world directly, by asking us to imagine things about it:

> In this approach, learning about a model is not a matter of discovering facts about an abstract or fictional model system; it is a matter of exploring the web of imaginings prescribed by a scientist's model description. For example, the Newtonian model description asks us to imagine that various assumptions hold of the sun and earth, such as that the force between them obeys Newton's law of gravitation. If we accept these initial assumptions, however, we are also to imagine that the earth moves in an ellipse, since this follows from the

> equation that we write down. This is simply part of the conventions that govern our interpretation of model descriptions. That the earth moves in an ellipse is therefore part of the content of the model, even though this was not specified in the model description.
>
> (459)

Whatever its other advantages (e.g. explaining scientific discourse), if we focus solely on the ability of models to provide information about empirical reality, the positing of a representation between us and reality by the indirect model seems problematic, not least because the mechanisms by which we get from the properties of the model to the real world properties the model supposedly represents is obscure. In order to see this, we must recognise a problem that each of these views share; a problem inherited from the Waltonian model on which they both rely.

The Waltonian framework of fictional engagement might seem to be initially appealing as an analogy for understanding SM, insofar as it offers certain plausible constraints on appropriate 'make-believe', constraints that govern, in part, the 'fictional' truths implied by/contained within SM. Yet, the lack of detail concerning the crucial elements of the view is, I think, sufficient to threaten the plausibility of the above accounts.

In both fictional accounts of SM we have a) a set of descriptions that give the bare content of the model, b) principles of generation that allow us to go beyond such descriptions to posit properties that will be true according to the model, and c) by some process of inference, perhaps resemblance in the case of indirect fictions, a 'mapping' from these properties to those of the real world system we are representing/imagining. To take an example from Toon's account: a model of the ideal pendulum is a description that prescribes us to imagine that the target, the real ball and spring system we have in front of us, is exactly as the text presents it: we have to imagine the spring as perfectly elastic and the bob as a point mass. So, SM are types of surrogative reasoning that allow us to learn about the target system via principles of generation that constrain what we are to legitimately imagine about the target system. But what are these principles and where do they come from? What justifies the choice of some rather than others?

In an attempt to answer these questions, Salis and Frigg (2017) contend that, to be successful, SM must be constrained at least by a reality principle (of the kind noted above) as well as a 'mutual belief' principle, a principle directed towards the mutual beliefs of the members of the community in which the story originates. They state:

> While these principles can be at work in certain TEs or SMs, other options may be possible. Meynell (2014: 4162–4163) points out that different kinds of TEs make use of different principles, and which ones are chosen depends on disciplinary conventions and interpretative

> practices. Specifically, she points out that 'which principles of generation a physicist brings most automatically to a TE will tend to reflect her beliefs about reality as well as the various theories and projects upon which she currently works' (ibid., 4163). For this reason, neither the reality principle nor the mutual belief principle are in any way privileged and different principles may be needed in specific domains of scientific enquiry.
>
> (22)

This, it must be admitted, is not very informative, but perhaps the variable nature of the contexts here makes further general context-independent specifications impossible. Nonetheless, it seems evident that in the case of SM, principles of generation must come from current (scientific) beliefs about how the world is—specifically those concerning the relevant domain of investigation—and the various background theoretical assumptions and explicit theories of which the experimenter is aware and which the experimenter thinks applicable to the model she is working with. Where else could they come from? Any given set of principles must, obviously, be chosen prior to the formulation of the SM. They are clearly not discovered in the model because, recall, anything discovered in the model is a *product* of prior principles and explicit descriptions. But now, it is very difficult to see what epistemic work could be being done by the imaginative engagement with the model, since the imaginings are constrained prior to such engagement. What could we possibly learn from the model that we didn't put into it in the first place? And even if it did seem to us that certain properties of some phenomenon were 'revealed' in engaging with the model, at best that is a contingent product of the principles we chose in advance. What if we had chosen different principles?

Now, one cannot pretend that these observations by themselves undermine the fictional view of SM, and as we will see, I will go on myself to argue that the imaginative engagement present in SM is of cognitive value. Yet they do, I contend, reveal that all of the work in explaining the appeal to fictional make-believe, and justifying its epistemic value in scientific reasoning has still to be done. The appeal to Waltonian games of make-believe by itself throws little light, if any, on the cognitive value of SM or the epistemic function of imagination therein.[2]

One might suspect that part of the problem here consists in the view of make-believe being offered. Salis and Frigg are explicit that the epistemically operative imaginative activity involved in SM is propositional imagining rather than sensory imagining. We 'entertain', rather than commit in a belief-like way to some proposition p or set of propositions that may or may not be accompanied by imagery, which itself plays no cognitive role in SM. If SM involve an imaginative engagement only with propositions—including descriptions, diagrams, and formulae—then the scepticism I just outlined seems inevitable, although it does perhaps still allow for SM to possess a surrogative role in scientific reasoning, just not one in which make-believe as such plays any interesting or substantive role.

So, in this light, it is imperative to ask: why the insistence on propositional imagining? Salis and Frigg (p. 17) argue that imagery is neither necessary nor sufficient to the outcome of scientific thought experiments. They state that "not all factors that matter to the successful performance of a TE seem to have sensory-like correlates. When considering Galileo's cavity we do not seem to have a perception-like representation of the cavity being frictionless or the lack of air-resistance." Imagery is not necessary since the cognitive work is done by the descriptions, concepts, and mathematical formulae of which the SM, qua SM, entirely consist:

> we cannot form a perception-like representation of the concept of force without having a theoretical definition, which is usually given in linguistic and formulaic symbols. Similarly, Malileo's SM assumes these concepts, but he also requires theoretical knowledge of Lagrangean mechanics, general principles and laws, mathematical abilities, and logical inferential abilities . . . We need to grasp the relevant concepts, with or without forming a mental image of the objects and transformations they stand in for.

At root here is, I think, a fairly familiar account of the uselessness of imagery, the thought that the relationship between a mental image and what it represents is determined by a description or interpretation, and not that of any resemblance between the image and its objects. That is to say, the intentionality of images is determined by the intention of the agent in forming them, and not by the intrinsic properties of the image as such. This feature of images explains what has been termed the 'multi-use' thesis, namely that the same image can stand in for many different imaginings (or imaginative projects), and that an imagining's success conditions do not—at least not solely—depend on the features possessed by the accompanying imagery. For example, suppose that my aim to imagine a chiliagon involves an image of a figure that does not literally have a thousand sides. The relatively impoverished nature of this image does not, it seems, undermine my imaginative project. Indeed I could also use the very same image to successfully imagine a hundred-sided figure.[3]

It is perhaps interesting to note the relationship between this claim and my objection to the principles of generation view above, that you only get out of an imagining, or image, what you put into it. Traditionally, many philosophers have held that images are essentially uninformative since their content is wholly determined by our intentions. From this fact follow other purportedly essential features of imagery, such as their inability to allow for genuine observation, in the sense that, unlike the objects of perception, we cannot discover new features or hidden aspects of the objects we imagine, nor are there any features of an imagined scene of which we are only peripherally aware. Images are, the thought goes, wholly sustained by attention and there is thus no equivalent in imagery of the centre and periphery of awareness that we have in the visual field. (See McGinn 2006). Imagining is in these ways essentially intentional, and in the context of gaining knowledge about the world, imagery is at best a merely psychological heuristic aid. As such, there is no obvious way in which it could play a genuine cognitive role in informing us about reality. We will return to these issues below.

In order to consider whether this view of imagery is accurate, it will be helpful to consider an issue that is not often discussed, and certainly has not been in the context of this topic: the transparency of imagery. I will argue that the non-transparency—or relative 'opacity'—of imagery is what allows it to play a significant cognitive role in scientific reasoning.

3 Imagery and Transparency

Given the above observations, it is difficult to see how visualisation itself could play a useful explanatory role in scientific reasoning. Rather, the role for visualisation is at best merely heuristic, a view supported by the purported fact that our imaginings are not directed at images themselves but at the 'things' the image is of. The intentional object of imagining, it seems, is not images. Images are merely, in the words of Kind, a type of 'mental paint' with which the imagination takes its intentional objects, the objects that are imagined. This is one sense, it might be held, in which images are *transparent* to the objects of imagining. We employ images to help us imagine but we—so to speak—see right through them to the things we are actually imagining. As such, it might seem, imagination in SM can successfully latch onto the physical world that is the main target of SM via images. Call this feature of images their 'transparency-to-the-world'. This clearly fits quite nicely with the Direct Fiction view, and it does not seem to pose a prima facie obstacle either to the Indirect View. But let us first look more closely at the notion of transparency.

There are various formulations of transparency, and the notion has traditionally played a central role in representational theories of perception. Here are two formulations of transparency adopted and adapted from Matthew Soteriou (2010):

Strong Transparency: introspection of one's perceptual experience reveals only the objects, qualities, and relations one is apparently perceptually aware of in having the experience.

Weak Transparency: when one introspectively attends to what it is like for one to be having a perceptual experience, it seems to one as though one can only do so by attending to the sorts of objects,

> qualities, and relations one is apparently perceptually aware of in having that experience.

It is, unfortunately, not obvious exactly how one can apply these ideas to imaginative experiences.

This is because imagination is phenomenologically unlike perception in being a threefold relation—attitude/experience, imagery, world (as imagined)—and this is so even where representationalist theories of perception posit representational content, because the conscious awareness of imagery in imaginative experience has no direct equivalent in perceptual experience. So, for example, it is not clear whether, if we transpose these views about transparency to the imagination, the 'objects' of which one is aware in imaginative experience are to be thought of as images, or the 'real' intentional objects of the imaginings (which we imagine via imagery). It is also not clear whether an imaginative experience that involves imagery is to count as an experience of the imagery or not.

If images are transparent in the sense outlined above, then imaginative experience must, one might think, be transparent to the objects (i.e. not images) that are imagined. But, of course, this seems wrong insofar as we are generally aware of our imagery when, for example, visualising, and such an awareness does not seem to depend on paying attention to the intentional objects of images, which would be required if images were strongly transparent to their objects. Would it be accurate to say that imaginative experiences are transparent to images, and then somehow transparent to the world via the transparency of imagery? I do not think so, since it does not seem to me that either type of transparency applies easily to imaginative experiences. This is for a number of reasons.

When we introspectively attend to our imaginative experiences, in the normal case we are aware of them as such—independent of any imagery that might accompany them—and this awareness necessarily involves an awareness of the essentially intentional, i.e. voluntary nature of the state we are in. Let us call this 'phenomenological voluntariness'. We are responsible for, and have some control over, the initiation, continuation, and content of the state. The type of attitude that counts as imagining is one that normally involves some intentional active *effort*. This phenomenology is very different from that involved, normally, in the *passive* state that is perception. So, there are phenomenological, and possibly structural features of an imaginative experience that we are normally aware of simply in virtue of having that experience.

This tells, at least, against imagining being strongly transparent. It seems to leave open the possibility that imagining is, however, weakly transparent since it might be argued that we are aware of these features only by attending to the images of which such experiences consist. Images themselves contain features that seem to be intrinsically linked to this voluntary phenomenology and that also tell against their supposed transparency-to-the-world, even if we acknowledge that imaginings in general are not intentionally directed at images.

Famously, images lack saturation—they do not completely fill in the visualised field in the way that our perceptual field is completely filled in, endlessly detailed and has no gaps in it. If you imagine, say, a landscape or your mother's face, the level of detail contained in the image seems almost entirely dependent on how much you deliberately add. Unless this effort is made, our imagery is often clearly much less detailed and 'filled in' than the equivalent perception would be. Moreover, there is no equivalent in an imagined scene to the (sense of) fullness we are aware of in visual perception. The imagined scene is 'gappy'. A related point: images lack determinacy. When imagining a tiger there may be no determinate number of stripes the tiger has, until and unless you deliberately count/add them in your mind. Either as the cause of consequence of such features, images are entirely attention-dependent in a way perception is not. As noted above, there is no equivalent of genuinely observing perceptual objects, of scanning the visual perceptual field, noticing things peripherally and coming to discover, for example, hidden aspects of three dimensional objects.[4]

So, one way of arguing for the weak transparency thesis is to hold that these features of imagery demonstrate that we are aware of the phenomenological voluntariness of our imaginative experiences only in so far as we are aware of these features of imagery. Or to put it slightly differently, we are generally always aware that we are imagining in virtue (in part) of our awareness of the fact that the object of our awareness is an image rather than a percept.[5]

In fact, however, I think this gets things the wrong way around. Phenomenological voluntariness is usually simply a basic datum of our (conscious) imaginative experiences, not one that we need to infer from features of imagery. Images lack determinacy and saturation *because* their nature and content depends wholly on our conscious effort and attention, in a way that is utterly unlike perceptual experience. The fact that we are normally aware that we are imagining, even in cases where no imagery is present, as in propositional imagining, should make this evident. It is phenomenological voluntariness which, I think, ensures that imaginative experiences are not even weakly transparent, whether the objects of such experiences are taken to be images or not.

In this light, it seems that only some type of strong transparency thesis for imaginative experience could accommodate the use of imagery on the fictional theories of SM we have examined. If visual imagery is entirely transparent or plays only a mediating role in SM, then we have a way of understanding how fictional views can safely ignore the role of imagery in SM, because images do not interfere with or pose an obstacle on the path from SM to world. Our imaginings can succeed in latching onto the world as it is imagined as being (either directly or via a representation).

Propositional imagining does all the cognitive work and is neither hindered nor helped by any imagery that may accompany it.

If imagining is not even weakly transparent, however, both fictional views would, it seems, be forced to explicitly deny not only that images can play any cognitive role in scientific reasoning, but that where images do feature, they may be detrimental to such reasoning, which is obviously a much stronger claim. The indirect fictional view will, for example, struggle to use a notion of visual resemblance to explain how we latch onto the world via SM, since images differ from their objects in all of the substantial ways discussed above. Relying on them for epistemic guidance would hence be extremely misleading. The direct view also seems forced to deny that any epistemically valid imagining of the world in an SM can rely in any way on imagery. But then it is also difficult to see on this view how we can avoid relying on imagery when we 'imagine of the world' where the phenomena we are trying to explain—e.g. the behaviour of gases—are understood in sensory terms. This is because insofar as the imagination is involved at all in SM it would necessarily be sensory, rather than merely propositional.

On the one hand, these stronger claims about imagery in SM are still compatible with the idea central to both fiction views that all the cognitively valuable work is done by propositional imagining. As such, they raise difficulties only for views that explicitly rely on visualisation to do serious epistemic work in scientific reasoning. On the other hand, however, they seem implausible when considering well-known examples of SM and TE where imagery appears to play a crucial and ineliminable role. Moreover, the essential opacity of imagination, I will argue, is central to understanding the important epistemic value that sensory imagining possesses in scientific reasoning.

4 Imagery, Aesthetic Feelings, and Understanding

Let us begin by noting that the multi-use thesis and essentially interpretive nature of imagery—when employed in the service of an imaginative project—do not suffice to undermine the potential cognitive significance of imagery. Imagery is clearly essential, for example, to some judgements of topological similarity, as numerous experiments have shown (see Gendler 2004).

More importantly, from our point of view, there is reason to think that imagery might possess important cognitive value arising from its connection with certain affective states that themselves possess cognitive import. Tamar Gendler (2004) has argued, citing Damasio's research, that imaginative rehearsal can bring us to new beliefs that may be unavailable to us if we reason in a disinterested purely hypothetical way. For example, she says, people who are afraid of public speaking or flying can repeatedly imagine themselves performing these activities until their fears are overcome. Note that although visual imagery is clearly at the forefront of this imagining, such a project might also involve proprioceptive and kinaesthetic and other viscerally oriented imagery—such as emotionally-laden memories—that draws on affective elements to increase the power of the visual imagery.

Turning to Mach's well-known thought experiment designed to show the force required to prevent an object sliding down a frictionless plane, she contends:

> Contemplation of an imaginary scenario (the cut string laid atop the prism) evokes certain quasi-sensory intuitions, and on the basis of these intuitions, we form a new belief about contingent features of the natural world (that the weight of four balls offsets the weight of three balls). This belief is produced not inferentially, but quasi-observationally: the presence of the mental image plays a crucial cognitive role in its formation.
>
> (1160)

It might seem that the role of such 'quasi-sensory' intuitions is, however, rather limited in nature and certainly cannot by itself offer the kind of justification that one might seek in scientific reasoning. I will return to this below. Yet for the moment we should note that even if this sort of imaginative reasoning serves a heuristic rather than surrogative function, that is a non-negligible source of some important epistemic value. Despite some warranted scepticism about the analogy between SM and fictions, Currie (2016) suggests that both enable us to see how a system operates under certain conditions: a gravitational system in the one case, a system of interacting persons in another. He cites Catherine Elgin, who argues that both involve simplifying certain patterns of causal interaction, and thereby "select and isolate, manipulating circumstances so that particular properties, patterns, connections, disparities and irregularities are brought to the fore" (303).

Propositional imaging here does a tremendous amount of work, of course, in setting up the relevant scenarios in imagination and driving the sensory imaginings that follow from it. But I contend that, at least in many cases, the relevant aspects only get highlighted as salient, patterns are recognised, and attention is only successfully focused by an engagement with sensory, imagistic imagining. This is primarily a result of the deep connection between imagery and aesthetic feelings and the epistemic function that such feelings serve in scientific reasoning. I take my cue here from Peter Kosso's (2002) observation:

> the hallmarks of scientific understanding are similar to an aesthetic feature associated with literature, music, and the visual arts. It is the feature described as coherence, harmony, and *inevitability of fit.*

> Aesthetics thus plays an epistemic role in science as an indication of understanding.
>
> (39)

I think it is this connection to aesthetic value, and particularly to aesthetic feelings of the kind Kosso mentions that we need to turn in order to understand the real value of imagery in scientific reasoning. I would like to claim that aesthetic judgements in science are plausibly understood as expressions of what I will call 'aesthetic-epistemic feelings' that serve a genuine cognitive and epistemic function. I will propose a naturalistic account of these feelings in terms of sub-personal processes of representing and assessing the relation between cognitive processes and certain properties of the stimuli at which they are directed.[6]

A number of psychologists have recently become interested in a range of phenomena that are usually referred to as epistemic feelings. That is, there are certain quite common affective conscious states, felt to be positively or negatively valenced, that arise in what can be broadly referred to as epistemic contexts. Drawing on a long list compiled by Arango-Muñoz and Michaelian (2014), two of the feelings relevant to our present topic would include, for example:

- **The feeling of knowing**: Koriat (2000) wrote that the 'feeling of knowing' is in some way a measure of the accessibility of the knowledge one has, correct or incorrect.
- **The feeling of understanding:** A feeling of intellectual satisfaction that motivates the endorsement of an explanation, a sense that we have achieved an understanding of a phenomenon that was not clearly understood before (Gopnik 1998; Trout 2002). This is sometimes called the "ah ha" feeling (Mangan 2001) or the Eureka feeling.[7]

The primary focus of the psychological literature has centred on the nature and function of such feelings, and whether they are in fact heuristically valuable in assessing the information they seem to concern. Although there is naturally much debate concerning both issues, there seems to be some consensus that epistemic feelings are the valenced experienced manifestation of some kind of quick and dirty, low-level, subpersonal mechanism that monitors the performance of different cognitive processes. There is evidence for neural correlates of epistemic feelings, and there is evidence that some epistemic feelings do play a justificatory role in, for example, accurately predicting future cognitive performance, and in acting as a stimulus to judgement.[8] In other words, my feeling of uncertainty or of knowing will incline me to (and arguably give me reason to) judge that I really am uncertain or that I do know.

Before looking further at the details, we can note that a few philosophers have also taken feelings seriously as playing an important role in cognition. For example, in discussing the so-called Frame Problem for AI, Ronald DeSousa (2008) argues that feelings and emotions serve to stop the potentially endless regress of deliberation by revealing patterns of saliency among objects of attention and strategies. And Christopher Hookway (2008) has argued that affective states in general embody evaluations that appeal to standards and principles of reasoning (for example, concerning inductive reasoning) that we may not be able to articulate. He claims that even if one can offer no reasons to justify one's acceptance of some inference, they may possess an *immediate phenomenology* of being compelling, as having epistemic salience, which is an affective state rather than a cognitive one (Cf. Elgin 2002; Thagard 2008).[9]

Let's begin with the well-studied feeling of knowing.[10] This is an experience we have whenever we are asked to recall a piece of information that seems to us to indicate that we possess the information without us (as yet) being able to recall the information. Psychological studies of memory retrieval suggest that subjects can accurately determine whether they are going to be able to recall the information by means of the feeling of knowing before they even try to recall it.[11]

As noted above, psychologists have proposed that the origin of epistemic feelings is low-level metacognition, which involves a sub-personal mechanism that monitors the performance of different cognitive processes. There is debate about how exactly to understand the nature of such a mechanism, but one plausible way is to think of it as a heuristic device that evaluates mental activity by reference to external conditions and/or properties of the relevant cognitive processing—such as fluency or familiarity—as well as by reference to some salient concepts and theories, as the conceptual nature of some e-feelings suggests.[12]

What, then, are such feelings actually about? Arguably, it seems that epistemic feelings have a dual intentional content; on the one hand, they are in some way about our own cognitive states—such as our capacity to know or remember something—and on the other hand, they are about the actual specific piece of information or object concerned. Phenomenologically, however, the main focus of our epistemic feelings will often be simply the relevant object or task at which our cognitive processes are directed, which explains why we attribute aesthetic features to it, rather than to our own mental states. This will be because, after all, the object or task is what grabs most or all of our attention. For, on the one hand, it takes some cognitive effort to attend to our own mental processes, while on the other hand, such attention will normally be seen as irrelevant to the task in hand, for example carrying out the implications of a SM.

In any case, the feeling of understanding clearly plays a central role in scientific reasoning, and particularly in the utilisation of SM. One of the crucial factors that appears to have some influence on the elicitation of this and other epistemic feelings is the fluency of the task, where fluency is, roughly, the experienced ease with which mental content is processed (R. Reber, Schwarz and Winkielman 2004). As Reber, Brun and Mitterndorfer (2008) note, many studies have demonstrated that stimuli processed with greater ease elicit more positive effect (R. Reber, Winkielman and Schwarz 1998; Whittlesea 1993; Winkielman, Halberstadt, Fazendeiro and Catty 2006), and that fluency plays a role in the rejection of theories that are difficult to understand (McColm 2007). Processing fluency increases either through former exposure, such as with stimulus repetition and associative learning, which renders stimuli familiar, or through stimulus features, such as simplicity and symmetry that facilitate perceptual processing. (See Reber, Brun and Mitterndorfer (2008) for further discussion.)

Reber et al. (2008) set out to study the role of fluency in mathematical reasoning and its relationship to certain properties of stimuli—such as simplicity or symmetry—that have also been implicated in experiences of beauty. They examined the use of symmetry as a cue for correctness in an arithmetic verification task by manipulating the symmetry of sets of dot pattern addition equations.[13]

They found that speeded decisions about the correctness of these equations led to higher endorsements for both correct and incorrect equations when the addend and sum dot patterns were symmetrical. That is, people without enough time to analyse the problem instead use heuristic cues—the presence of symmetry—in their assessment of the correctness of a proposed solution. Therefore, they argue, this effect does not seem to be due to the fact that symmetry facilitates calculation or estimation.[14]

We ought, however, to proceed with some caution. Currie (2016), for example, has pointed out that in cases where people are asked about the truthfulness of a story their 'feeling for truth' is highly unreliable, often having more to do with, for example, the presence or rhyme, the vividness of the imagery, intensity of emotional stimuli, or the attractiveness of the speaker. Perceptual fluency—perceived ease of processing—is clearly operative here. Currie goes on to argue that, in contrast to fiction, SM are "not dependent for their value in learning on any particular formulation; rather they depend on their capacity to get good predictive or explanatory results or to achieve some other epistemic aim" (305).

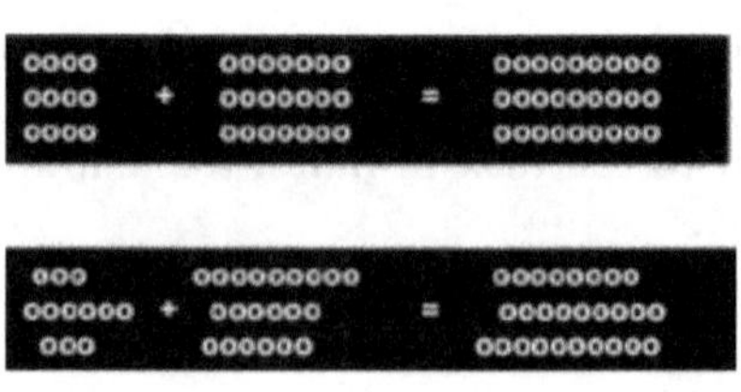

Figure 4.1

It is here that I think Currie is mistaken. First, because the formulation of SM is crucial to their epistemic import, and in particular to their explanatory value, which is in large part a function of the feeling of understanding that they can produce. It is important to note that the studies on fluency outlined above, even though they focus on mathematical judgements, are primarily concerned with perceptual processing. It is the connection between the 'perception' of imagery and the feeling of understanding that accompanies them, that is fundamental to the success conditions of at least some SM. Second, in this light, the relevant epistemic feelings at play in SM—particularly the feeling of understanding—are not valuable simply in terms of their reliability as 'truth-trackers'. The notion of understanding is more pertinent to the cognitive function of SM than truth.

In an earlier study Reber et al. (2004) also identified processing fluency at work in purported judgements of beauty, thus promising to connect judgements of truth and judgements of beauty together via the medium of fluency and via epistemic feelings. Specifically, Reber et al. appealed to previous empirical experiments as well as their own studies to argue that "aesthetic pleasure is a function of the perceiver's processing dynamics: The more fluently perceivers can process an object, the more positive their aesthetic response" (p. 364). They examined a number of variables taken to influence aesthetic judgements, such as figural goodness, figure-ground contrast and clarity, stimulus repetition and familiarity, symmetry, and prototypicality, and tried to trace their effects to changes in processing fluency. They also looked at the role of other variables, like visual or semantic priming, that appear to influence judgements of aesthetic pleasure. The range of studies cited and performed is too large to discuss in any length here but, in their own words, the main findings were the following:

> First, objects differ in the fluency with which they can be processed. Features that facilitate fluent processing include all the core features identified in objectivist theories of beauty, like goodness of form, symmetry, and figure-ground contrast, as well as variables that have not received attention in traditional theories of aesthetic pleasure, like perceptual and conceptual priming procedures. Second, processing fluency is itself hedonically marked and high fluency is subjectively experienced as positive, as indicated by psychophysiological findings. Third, the affective response elicited by processing fluency feeds into judgments of aesthetic appreciation, unless the informational value of the experience is called into question. Finally, the impact of fluency is moderated by expectations and attribution.
>
> (377)

Some features of their study and related studies are important to note. First, the stimuli for many of the experiments—with the exception of priming experiments—are often simple visual figures, rather than sophisticated visual artworks, and accordingly preference is given to visual stimuli. Second, the experiments depend upon the self-reports of subjects concerning their affective responses, including judgements of pleasure and preference, but rarely of beauty, as we'll note below. Third, the authors claim that although aesthetic preferences *always* depend on fluency, experiences of fluency can be influenced by both biological equipment and socialisation. This is important in explaining the differences between, for example, novice and expert preferences and a number of other variables resulting from the experiments. This finding fits too with other research in showing that epistemic-aesthetic feelings seem to be cognitively penetrable: that is, they can be influenced by top-down processing involving certain background knowledge, assumptions and expectations. For example, complexity may sometimes be preferred to simplicity by experts because it facilitates access to the meaning of the stimulus. "That is, a decrease in perceptual fluency due to complexity may be outweighed by an increase in conceptual fluency due to meaningfulness" (373).

One might wonder here how, if fluency processing can be the result of certain contingent features of one's background knowledge, expectations and so on, such feelings could possibly be relied upon for epistemic insight. Such a worry, I believe, is misplaced. This is because, once we have the notion of expertise in play—scientific or artistic—we have a grasp of certain conditions that appropriately govern the first-order responses of the expert and this ensures that such responses possess a measure of objectivity. This is the case even where, as in art, the relevant expertise is itself in part the result of 'mere' historical conventions, but in the case of scientific expertise the role of mere convention is clearly subordinate to more or less universally accepted methodological constraints and well-established theories.[15]

I have argued elsewhere that, in light of these findings and other considerations, we can think of (some) epistemic feelings as possessing aesthetic attributes, or of (some) aesthetic feelings as exhibiting epistemic attributes, or of a range of feelings that are jointly aesthetic-epistemic in nature (Todd 2017). First, epistemic feelings, like aesthetic experiences, are valenced. They are experienced as positive or negative. This, plausibly, is a result of them being manifestations of some sort of metacognitive monitoring of how well or badly, efficiently or inefficiently, our cognitive process are engaging with an object or task. Second, epistemic feelings, like aesthetic experiences, can be and perhaps often are quick and dirty responses that are opaque, at least initially, to the reasons that ground them. Third, let's not forget, in the context of scientific practice, these epistemic feelings are often characterised as aesthetic. The nature of the experiences is, to the subjects undergoing them, experienced as being of the same (rough and ready) kind as those present in paradigmatic aesthetic contexts.[16] Finally, as we have seen, many of the criteria and stimulus properties occur as the objects and causes of both aesthetic and epistemic experiences—symmetry, simplicity, fluency, order, clarity, and so on.

The two notions I want to concentrate on here are those of 'understanding' and 'fit'. The 'Aha!' moment that encompasses a surprising discovery, or simply reflects the fact that some impressive intellectual achievement or insight has been accomplished can be the result of a feeling of fluency, and seems often to have something to do with the notions of harmony or fit. Here we find, as many have pointed out, some striking continuities between aesthetic judgement and scientific reasoning, in terms of the appreciation of patterns, connections, symmetries, and harmonies common to each.[17] This notion of fit, as Wittgenstein long ago remarked, seems central to many aesthetic experiences, although it is very difficult to articulate precisely. Yet I think that there is a distinctive type of feeling that accompanies the 'perception' of fittingness and it is very closely related to, or perhaps a variant of the feeling of understanding.

Mike Stuart (2018) lists a number of ways that thought experiments in science can increase understanding, which he divides into the sub-categories of explanatory understanding, practical understanding, and objectual understanding (OU). If we look only at the latter, Stuart characterises this in terms of Camp's notions of perspectives, characterisation, and frames. OU-oriented thought experiments, he states, can be understood as frames which are meant to lead us into certain characterisations. He says:

> Some of Darwin's opponents characterized the eye as a watch. Even if we saw a watch on a deserted island, we would nevertheless assume the watch was created because of its complexity and obvious purposefulness. This frame suggests a characterization of the eye as being like a watch, which is complex, purposeful and the product of intentional creation. That is how the frame casts doubt on the idea that eyes are the result of a series of chance mutations. Darwin, on the other hand, presents a competing characterization using his thought experiment, which narrates a series of mutations that could plausibly result in a fully functioning eye beginning with a single nerve. This characterization makes it easy to see the eye as having evolved.
>
> (535)

This kind of understanding is not fundamentally propositional or even factive, and nor is much of the understanding that results from scientific reasoning, as for example Elgin has frequently pointed out in her work (e.g. 2002, 2007, 2014). As she suggests, the idealisations that abound in science are essential in offering an irreplaceably direct epistemic to certain features of the phenomenon we want to understand. Imagery can play a crucial role in doing so, and insofar as SM employ imagery to highlight certain properties, or to focus our attention in certain ways, the understanding that results is partly non-propositional and affective in nature. Such understanding, I contend, frequently has an aesthetic character, a feeling of fit.

One key component to the feeling of understanding present in scientific reasoning is unification. As philosophers such as De Regt, Elgin, and Stuart have pointed out, understanding a phenomenon in science—such as the kinetic theory of gases explained in Boyle's Law—frequently requires unifying often disparate elements together, seeing how they interconnect in a way that is intelligible. Such unification is often aided by visualisation, no doubt in part because human beings are primarily visual creatures. This is why we rely on visualisation for understanding even the most abstract areas of science, and why, when phenomena outstrip our ability to visualise them—such as in the case of various puzzling quantum phenomena—it is difficult to grasp them in a way that satisfies our desire for explanation and understanding.

Importantly for our purposes, it is not just the role of visualisation as such that plays a central role in understanding, but also the connection that it has to certain affective-epistemic states. We saw earlier that various philosophers have appealed to the cognitive role affective states may play in guiding our judgements in lieu of propositional or 'rational' processes, and we have seen how this may be explained in part by processing fluency and the epistemic-aesthetic feelings that are manifestations of it. De Regt and Dieks talk of how understanding in science is often the result of certain intuitive 'judgements' we make, intuitions that often but not always arise out of visualisation and which play a crucial heuristic role. A scientific theory, they contend, should be intelligible, in that we want to be able to 'grasp how the predictions are generated, and to develop a feeling for the consequences the theory has in concrete situations' (144). They note that, in the visualised case:

> A simple illustration is the use of 'field lines' in electrostatics . . . Although intuitive application of this concept is possible only in simple situations, it is quite useful to get a feeling of how electrostatic systems behave. And this . . . is precisely what it means to have physical understanding of the situation in question: "if we have a way of knowing what should happen in given circumstances without actually solving the equations, then we 'understand' the equations, as applied to these circumstances.[18]
>
> (159)

The role of intuition has also been noted by Cartwright (1983), who argues that applying a model to a real system is a matter of intricate approximation. Formal principles, telling us how to get from a theory via a model to a description of a real system, do not exist: "There are just rules of thumb, good sense, and, ultimately, the requirement that the equation we end up with must do the job" (ibid.: 133). I suggest that 'good sense' here is nothing other than the kind of intuitive epistemic-aesthetic feeling of understanding or fit that I have been discussing.

To sum up briefly: several philosophers have noted that visualisation is crucial to some scientific models, and hence can play a fundamental role in scientific reasoning. Others have contended that the notion of understanding rather than truth is central to the cognitive value of SM, and several have discussed the role of feelings, intuitions, and aesthetic judgements in science. Looking closely at experimental work on processing fluency, and on the epistemic role of certain aesthetic feelings allows us, I have argued, to understand how all of these aspects may be connected, and hence to gain a clearer overview of how scientific reasoning works.

It remains only to end with some remarks on the connection between imagistic opacity and the relevant affective states, as I promised at the very beginning. Unfortunately, however, it is rather difficult to be precise here, and I can't pretend to offer anything other than the briefest sketch of what are, after all, certain cognitive processes that operate largely at sub-personal levels.

Opacity, recall, is that feature of images in virtue of which we are aware of the 'phenomenological voluntariness' of the mental state we are in: roughly, of the fact that we are imagining and are therefore normally responsible for the content of our imagery. It is partly this feature of imagining and imagery that leads directly to scepticism about the epistemic value of each in scientific reasoning. We can now begin to see, however, how imagery can be driven—and can in turn drive—certain affective reactions (aesthetic-epistemic feelings and intuitions) that constitute the feeling of understanding.

To be sure, these feelings may in certain cases arise independently of imagery, as is arguably the case in abstract mathematics. Nonetheless, where imagery seems inescapable in scientific reasoning, it is often accompanied by feelings that stem directly from the phenomenal character of imagining itself. That is, where imaginings are driven by processes that underpin the kinds of epistemic-affective feelings we experience in scientific reasoning, the feeling of understanding or fit that results is not the kind of feeling that could result from some passive perceptual or belief state. The conscious effort, intention, and attention that goes into imagining 'seeks' a kind of reward or satisfaction that only such aesthetic-epistemic feelings can provide—or rather, of which such feelings are the natural manifestation. Where visualisation is present, such feelings will necessarily be bound up with the kind of visual aesthetic features—of harmony, coherence, unity, symmetry, and so forth—that we have discussed.

As such, what and how we visualise in SM will be driven in part by theoretical principles and considerations, and in part by propositional imaginings, but also by affective intuitions that may be below conscious awareness. In turn, our visualised scenarios may themselves arouse further affective states, and amongst such states the feeling of understanding is one of the most important in scientific reasoning.

Acknowledgements

I would like to thank the editors very much for their extreme patience and excellent feedback on this chapter.

Notes

1. See also Salis and Frigg (2017: 2).
2. Cf. Currie (2016) for a similar sceptical view about the analogy between SM and fiction, though he appeals to different considerations.
3. For a good overview of these issues see Gregory (2016).
4. For an accessible discussion of all of these features see McGinn (2006).
5. The Perky experiments might seem to cast doubt on this claim, but their interpretation is controversial, and in any case the phenomena they show are applicable only in very restricted settings.
6. What follows draws largely on my paper, Todd (2017).
7. Note here that I am concerned primarily with the phenomenology of understanding and hence the role of feeling in SM. There are, of course, other ways of interpreting the role of 'understanding' in scientific reasoning, some of which I will discuss below (e.g. De Regt and Dieks 2005; Stuart 2018).
8. See Arango-Muñoz and Michaelian (2014) and Arango-Muñoz (2014) for discussion.
9. These lines of thought are compatible with recent philosophical accounts of the emotions as states or processes whose function is largely to disclose certain evaluative states of affairs, and which is accomplished primarily via their experiential phenomenology. See Deonna and Teroni (2012) for discussion.
10. The *locus classicus* is Koriat (2000).
11. See Reder (1987).
12. See Arango-Muñoz (2014). One recent study has shown that certain gut feelings, such as the feeling of error, can be reliable indicators of the accuracy of one's own mental performances in mathematical reasoning tasks. Arango-Muñoz et al. (2015) showed, for example, that the feeling of error was strongly correlated with arithmetic errors in a Number Bisection Task, where the instructions and time restriction imposed ensured that the answers to the questions were the first that quickly and intuitively came to participants' awareness. This suggests, in the words of the authors that "this type of feeling-based metacognition provided participants with accurate assessments of their ongoing cognitive processes without the necessity of effortful and analytical thinking" (31).
13. Participants were presented with dot pattern addition equations, one by one. Half were correct (e.g., 15 □ 18 □ 33), and half were wrong (e.g., 15 □ 18 □ 27). The incorrect sums were either smaller or greater than the corresponding correct result, and the difference between the two was balanced across symmetry conditions. Each equation was shown twice, once as a symmetric pattern, once as an asymmetric pattern, yielding 96 dot-pattern-shaped equations (Figure 1). All symmetric patterns were rectangles, with from three to five rows. Operands with asymmetric patterns always had as many dots

and as many rows as the same operand with symmetric patterns, but the dots were rearranged so that they possessed neither vertical nor horizontal symmetry.

14. Semir Zeki, Romaya, Benincasa and Atiyah (2014) have directly tackled mathematical experiences of beauty, focusing on a list of mathematical formulae. Zeki et al. found some agreement about the aesthetic judgements of certain formulae. They found, for example, that the formulas most consistently rated as beautiful were both before and during the fMRI scans were Euler's identity ($1 + e^{i\pi} = 0$) and that the one most consistently rated as ugly was Ramanujan's infinite series for $1/\pi$. The experimenters were focusing on the hypothesis that such experiences correlate with activity in the same part of the 'emotional brain'—specifically A1 of the Orbito Medial Frontal Cortex (oMFC)—as experiences of beauty towards other paradigmatic aesthetic stimuli, such as visual artworks and music. The results in favour of a positive conclusion come with the observation that this area is also active in a variety of (arguably non-aesthetic) conditions, including experiences relating to pleasure, reward and hedonic states. Although this might be interpreted as showing that similar brain areas are operative in aesthetic experiences and experiences of understanding, I think we need to be extremely cautious in investigating what sort of overlap is implied. For example, as a certain level of understanding is required to have the aesthetic experiences in the first place, but I want to argue that the feeling of understanding can itself be an aesthetic feeling.
15. For further discussion see Walton (1970) and De Regt and Dieks (2005).
16. For example, in response to Ramanujan's work on identities, Watson said that it gave him "a thrill which is indistinguishable from the thrill which I feel when I enter the Sagrestia Nuova of Cappella Medicee and see before me the austere beauty of 'Day', 'Night', 'Evening' and 'Dawn' which Michelangelo has set over the tombs of Giuliano d' Medici and Lorenzo d' Medicii" (Quoted in Chandrasekhar 1987: 61).
17. Cf. Root-Bernstein (2002), Lipton (2004: Ch. 9), Arnheim (1996).
18. In the non-visualised case: "many theoretical physicists have developed a familiarity with, and intuition for, the general behaviour of the solutions of the mathematical equations they use. This enables them to acquire a feeling for the qualitative behaviour of the described systems without invoking picturable physical mechanisms. For instance, it is possible to get an intuitive feeling for how quantum-mechanical systems in two-slit-like situations behave, by familiarity with the linear character of the Schrödinger equation" (160).

Bibliography

Arango-Muñoz, S. (2014), 'The Nature of Epistemic Feelings', *Philosophical Psychology*, 27: 193–211.

Arango-Muñoz, S. and Michaelian, K. (2014), 'Epistemic Feelings, Epistemic Emotions: Review and Introduction to the Focus Section', *Philosophical Inquiry* II: 97–122.

Arango-Muñoz, S. and Volz, K. (2015), 'Oops, Scratch that! Monitoring One's Own Errors during Mental Calculation', *Cognition*, 146: 110–120.

Arnheim, R. (1996), 'Beauty as Suitability', *Journal of Aesthetics and Art Criticism*, 54: 251–253.

Cartwright, N. (1983), *How the Laws of Physics Lie*. Clarendon Press.

Chandrasekhar, S. (1987), *Truth and Beauty: Aesthetics and Motivations in Science*. University of Chicago Press.

Currie, G. (2016), 'Models and Fiction, Fictions as Models', *The Monist*, 99: 296–310.

Deonna, J. and Teroni, F. (2012), *The Emotions: A Philosophical Introduction*. Routledge.

De Regt, H. and Dieks, D. (2005), 'A Contextual Approach to Scientific Understanding', *Synthese*, 144: 137–170.

De Sousa, R. (2008), 'Epistemic Emotions', in G. Brun, U. Doguoglu and D. Kuenzle (eds.), *Epistemology and Emotions*. Ashgate.

Dokic, J. (2012), 'Seeds of Self-Knowledge: Noetic Feelings and Metacognition', in M. J. Beran, J. Brandl, J. Perner and J. Proust (eds.), *Foundations of Metacognition*. Oxford University Press, pp. 302–321.

Elgin, C. Z. (2002), 'Art in the Advancement of Understanding', *American Philosophical Quarterly*, 39: 1–12.

Elgin, C. Z. (2007), 'Understanding and the Facts', *Philosophical Studies*, 132: 33–42.

Elgin, C. Z. (2014), 'Fiction as thought experiment', *Perspectives on Science*, 22: 221–241.

Engler, G. (1990), 'Aesthetics in Science and Art', *British Journal of Aesthetics*, 30: 24–33.

Frigg, R. (2010a), 'Models and Fiction', *Synthese*, 172: 251–268.

Frigg, R. (2010b), 'Fiction and Scientific Representation', in R. Frigg and M. Hunter (eds.), *Beyond Mimesis and Convention: Representation in Art and Science*. Springer.

Gendler, T. (2004), 'Thought Experiments Rethought—and Reperceived', *Philosophy of Science*, 71: 1152–1163.

Glynn, I. (2010), *Elegance in Science: The Beauty of Simplicity*. Oxford University Press.

Gregory, D. (2016), 'Imagination and Mental Imagery', in A. Kind (ed.), *The Routledge Handbook of Philosophy of Imagination*. Routledge, pp. 97–109.

Gopnik, A. (1998), 'Explanation as Orgasm', *Minds and Machines: Journal for Artificial Intelligence, Philosophy and Cognitive Science*, 8: 101–118.

Hookway, C. (2008), 'Epistemic Immediacy, Doubt and Anxiety: On a Role for Affective States in Epistemic Evaluation', in G. Brun, U. Doguoglu and D. Kuenzle (eds.), *Epistemology and Emotions*. Ashgate.

Kivy, P. (1991), 'Science and Aesthetic Appreciation', *Midwest Studies in Philosophy*, 16: 180–195.

Koriat, A. (2000), 'The Feeling of Knowing: Some Metatheoretical Implications for Consciousness and Control', *Consciousness and Cognition*, 2: 149–171.

Kosso, P. (2002), 'The Omniscienter: Beauty and Scientific Understanding', *International Studies in the Philosophy of Science*, 16: 39–48.

Lipton, P. (2004), *Inference to the Best Explanation*. Routledge.

Mangan, B. (2001), 'Sensation's Ghost: The Non-Sensory "Fringe" of Consciousness', *Psyche* 7: 1–18.

McColm, G. (2007), 'A Metaphor for Mathematics Education', *Notices of the American Mathematical Society* 54: 499–502.

McGinn, C. (2006), *Mindsight*. Harvard University Press.

Meynell, L. (2014), 'Imagination and Insight: A New Account of the Content of Thought Experiments', *Synthese* 191: 4149–4168.

Osborne, H. (1984), 'Mathematical Beauty and Physical Science', *British Journal of Aesthetics*, 24: 291–300.

Reber, R., Brun, M. and Mitterndorfer, K. (2008), 'The Use of Heuristics in Intuitive Mathematical Judgement', *Psychonomic Bulletin & Review*, 15: 1174–1178.

Reber, R., Schwarz, N. and Winkielman, P. (2004), 'Processing Fluency and Aesthetic Pleasure: Is Beauty in the Perceiver's Processing Experience?', *Personality and Social Psychology Review*, 8: 364–382.

Reber, R., Winkielman, P. and Schwarz, N. (1998), 'Effects of Perceptual Fluency on Affective Judgments', *Psychological Science*, 9: 45–48.

Reder, L. M. (1987), 'Strategy Selection in Question Answering', *Cognitive Psychology*, 19: 90–137.

Root-Bernstein, R. (2002), 'Aesthetic Cognition', *International Studies in the Philosophy of Science*, 16: 61–77.

Salis, F. and Frigg, R. (2017), 'Capturing the Scientific Imagination', in P. Godfrey-Smith and A. Levy (eds.), *The Scientific Imagination*. Oxford University Press.

Soteriou, M. (2010), 'Perceiving Events', *Philosophical Explorations*, 13: 223–241.

Stuart, M. (2018), 'How Thought Experiments Increase Understanding', in M. Stuart, Y. Fehige and J. Brown (eds.), *The Routledge Companion to Thought Experiments*. Routledge, pp. 526–544.

Thagard, P. (2008), *Hot Thought*. Bradford Books.

Todd, C. (2008), 'Unmasking the Truth Beneath the Beauty: Why the Supposed Aesthetic Judgements Made in Science May Not Be Aesthetic At All', *International Studies in the Philosophy of Science*, 11: 61–79.

Todd, C. (2017), 'Fitting Feeling and Elegant Proofs: On the Psychology of Aesthetic Evaluation in Mathematics', *Philosophia Mathematica*, 26: 211–233.

Toon, A. (2016), 'Imagination in Scientific Modelling', in A. Kind (ed.), *The Routledge Handbook of the Philosophy of Imagination*. Routledge.

Trout, J. D. (2002), 'Scientific Explanation and the Sense of Understanding', *Philosophy of Science*, 69: 212–233.

Walton, K. (1970), 'Categories of Art', *Philosophical Review*, 79: 334–367.

Whittlesea, B. W. A. (1993), 'Illusions of Familiarity', *Journal of Experimental Psychology: Learning, Memory, & Cognition* 19: 1235–1253.

Winkielman, P., Halberstadt, J., Fazendeiro, T., and Catty, S. (2006), 'Prototypes Are Attractive because They Are Easy on the Mind', *Psychological Science*, 17: 799–806.

Zeki, S., Romaya, P., Benincasa, D. and Atiyah, M. (2014), 'The Experience of Mathematical Beauty and Its Neural Correlates', *Frontiers in Human Neuroscience*, 8: 1–12.

5 Beauty, Truth and Understanding

Milena Ivanova

1 Introduction

Many scientific theories have been praised for their aesthetic qualities. Newtonian mechanics and Einstein's theory of relativity are given as examples of a beautiful theory. The beauty of scientific theories is often used in the evaluation of their likelihood of being true or in the estimation of their expected empirical success. That is, often scientists place epistemic import on the aesthetic values of theories, deciding whether to commit to a theory in light of its aesthetic appeal, especially in situations when sufficient empirical data is not available to guide such a decision. The question then arises whether we can trust aesthetic considerations to be playing an epistemic role in science and informing our attitudes towards scientific theories.

In this chapter I outline accounts that have defended the epistemic role for beauty and aesthetic values in science, claiming that there is a link between an aesthetically appealing theory and its likelihood to be true. After challenging the plausibility of these accounts, I turn to an alternative defence for the relevance and importance of aesthetic considerations in science. It is argued that science has many goals, truth and empirical success being the usual favourites, but it also aims at offering understanding of phenomena and such understanding can be achieved in the absence of truth. By focusing on the concept of understanding, I argue that aesthetic factors are intricately linked to our own cognitive make up and desire to understand the world around us, shaping our inferential patterns and guiding the construction and acceptance of scientific theories.

2 Beauty in the Practice of Science

Just like many different forms of artwork can be praised as beautiful or aesthetically appealing, in science many different objects are claimed to be beautiful or possess aesthetic properties. We find beauty in the phenomena, such as solar eclipses, aurora borealis, diffracted light rays, pictures of oscillating particles, scientific models like the structure of

DNA molecules, mathematical proofs like Euclid's Elements, experiments like Rutherford's explorations of uranium radiation, and, of course, scientific theories. In this chapter, my focus is primarily on the aesthetic properties of scientific theories. In what way have theories been considered beautiful? There is certainly a plurality of ways in which a theory can be beautiful. A theory can be simple, elegant, unified, obey symmetry principles, be consistent and coherent. Theories can exemplify these aesthetic qualities or virtues in multiple ways. For instance, Newtonian mechanics can be regarded as simple because it can describe the motion of bodies with three rather simple laws of motion and a law of gravitation featuring few parameters. The general theory of relativity is certainly more complex than Newton's theory of gravity, the number of equations is larger and there are more parameters in them. But the theory offers a much more economical way of understanding the concepts involved in describing motion and forces by unifying gravity and inertial mass, the concepts of space and time into a space–time continuum, so in this way it provides us with simpler and more unified understanding of gravitational phenomena. Mendeleev's periodic table elegantly classifies all 117 elements by their common property of atomic number.

Beauty is often seen as a motivator in scientific enquiry, but is also given significant epistemic weight by being regarded as an indicator of the theory's truth. When it comes to its motivational role, scientists often claim that they study nature in order to find the beauty within it. During the heated debates in France at the turn of the twentieth century regarding the aim of science and scientific progress, aiming to oppose science scepticism motivated by the argument from the 'bankruptcy of science', Henri Poincaré (2001) offers a defence of science appealing to its beauty. Poincaré argues that we cannot justify the idea that science is valuable in its own right, since we need to take practical constraints into account, but we cannot regard science as valuable only insofar as it delivers us products, such as technological advances. Science is valuable, he argues, because in constructing scientific theories and uncovering underlying relations in the phenomena, we experience an aesthetic response. As Poincaré states, "[t]he scientist does not study nature because it is useful to do so. He studies it because he takes pleasure in it, and he takes pleasure in it because it is beautiful" (Poincaré 2001: 368). Poincaré clarifies how he understands this concept of beauty: "I am not speaking, of course, of the beauty which strikes the senses, of the beauty of qualities and appearances. I am far from despising this, but it has nothing to do with science. What I mean is that more intimate beauty which comes from the harmonious order of its parts, and which pure intelligence can grasp" (ibid.). In similar fashion, the Nobel laureate Subrahmanyan Chandrasekhar also claims that "in the arts as in the sciences, the quest is for the very same elusive quality: beauty" (Chandrasekhar 1987: 52).

Many scientists have appealed to the concept of beauty when providing justification of their commitment to a theory. Often, the aesthetic features of a theory, such as simplicity and unity, are seen as convenient considerations in theory choice. As Ernst Mach (1984) argues, we should always aim to explain the phenomena in the most conceptually economical manner, because simple theories are easier to use. He argues that science should be seen as an economical description of our observations, with simplicity being a guiding principle in the construction and evaluation of scientific hypotheses. In the *Science of Mechanics*, Mach argues that: "The aim of science is to be an 'economy of thought', to offer us a simple and concise classification of the observable appearances and enable the prediction of phenomena" (Mach 1984: 577). In a similar manner, Poincaré argues that there is a relationship between utility and simplicity, in that a simpler theory is easier to use, making simplicity a regulative principle in the construction and choice of theories. According to him,

> care for the beautiful leads us to the same selection [of theories] as care for the useful. Similarly economy of thought, that economy of effort which, according to Mach, is the constant tendency of science, is a source of beauty as well as a practical advantage.
>
> (Poincaré 2001: 369)

In addition to its heuristic role, beauty is often taken to stand in a special epistemic relationship to truth. Many scientists hold that a beautiful theory is more likely to be true, so if faced with a choice between two theories, the simplest or more beautiful theory is to be epistemically privileged. Paul Dirac is often stated to have adopted such an epistemic role for beauty in science, claiming that "one has a great confidence in [a] theory arising from its great beauty, quite independently of its detailed successes" (Dirac 1980: 40). Dirac takes beauty to be an indicator of the theory's truth, such that we can be confident in the truth of a beautiful theory even before the theory is supported by the empirical data. He claims that "[o]ne has an overpowering belief that [the theory's] foundations must be correct quite independently of its agreement with observation" (ibid.). Dirac claims that one had good ground to believe in the truth of the general theory of relativity prior to Arthur Eddington's expeditions, which were regarded as its first empirical confirmation, on the grounds of the theory's beauty. This conviction was shared by many physicists at the time, including Eddington himself as well as Werner Heisenberg, who argued that "[i]f nature leads us to mathematical forms of great simplicity and beauty we cannot help thinking that they are 'true', that they reveal a genuine feature of nature" (Heisenberg 1971: 68).

Chandrasekhar similarly argues that it is reasonable to believe that "we have evidence that a theory developed by a scientist, with an exceptionally well-developed aesthetic sensibility, can turn out to be true even if, at the time of its formulation, it appeared not to be so" (Chandrasekhar 1987: 64). In his 1979 lecture 'Beauty and the Quest for Beauty in Science' presented at the Fermi National Accelerator Laboratory, Chandrasekhar argues that scientists aim to uncover the beauty in nature and this is their primary motivation in science. He articulates an account of beauty, drawing from Francis Bacon's thoughts that beauty implies some 'strangeness in the proportion', understood as unexpected wonder or surprise, and Heisenberg's definition of beauty as 'conformity of the parts to one another and to the whole'[1] (ibid.: 70). Drawing upon this definition, Chandrasekhar goes on to explain in detail why Einstein's theory of general relativity is beautiful, achieving the unification of our fundamental concepts of space and time, and the concepts of matter and motion with "unerring sense for mathematical elegance and simplicity" (ibid.: 71). He argues that the discovery of beauty in nature is the most significant of achievements, that it is an "incredible fact that a discovery motivated by a search after the beautiful in mathematics should find its exact replica in Nature" (ibid.: 54).[2]

The employment of beauty is particularly interesting in the context of contemporary particle physics, with the research being driven to discover 'susy' (super symmetric particles) and the symmetry principle becoming something of an imperative in the community. The Nobel laureate Marry Gell-Mann, whose use of symmetry principles led to the advancement of the standard model and the discovery of the previously unknown particle now called omega minus, has also argued for the deep connection between beauty and truth (Hossenfelder 2018: 37). But while many remain enthusiastic about the fertility of the symmetry principle and building more colliders to discover susy particles, others are on the fence as to whether this principle should be trusted so much in contemporary physics and suggest beauty might be a systematic bias in contemporary physics, leading physicists to pursue research programmes that are not fruitful (Hossenfelder 2018).

3 Can Beauty Tell Us About the Truth of Our Theories?

The previous section showed that scientists do use aesthetic considerations in their decision-making and in deciding whether to trust a theory. This fact is particularly prevalent in contemporary high-energy physics, with some arguing that beauty claims have misled the physics community into fruitless projects. So should we trust beauty to be an indicator of truth? What grounds could we have to place trust in non-empirical factors such as aesthetic values of a theory? Traditionally the link between beauty and truth has been developed in both the rationalist and empiricist tradition. On the one hand, we can argue that beautiful theories correctly capture facts about the world and their aesthetic qualities latch on to beauty in the world. On the other hand, one can place confidence in a beautiful theory simply because, inductively, beautiful theories seem to have had a good track record of empirical success. In this section I outline both defences and focus primarily on the empiricist approach to defending the link between beauty and empirical adequacy or truth, as developed in the work of James McAllister (1989, 1996).

While rationalist defences of beauty are not particularly popular in the current philosophical literature, they have been employed since antiquity in an attempt to show that there is an intrinsic relationship between beauty and truth, and are implicitly assumed by many scientists. In the *Philosophiæ Naturalis Principia Mathematica* Isaac Newton proposed several methodological rules in science, one of which is to avoid superfluous causes in describing nature. The justification for this principle is that nature is itself simple. Such arguments are taken to be circular: our confidence in the claim that nature is beautiful, for instance for its simplicity, is justified upon reflection on our current best theories, which in their turn are meant to be true in light of their reflection of the beauty or simplicity of the world. It is also problematic to assume the beauty of nature a priori, since reflecting upon some of our theories can also lead us to make the contrary conclusion. For instance, if we take the standard model, often claimed to be an ugly and inelegant theory, to be reflecting the nature of reality, we would perhaps have to conclude that the world is not simple or unified as we would like it to be, and perhaps there is more complexity and irregularity in the world. I now turn to discussing empiricist views of the role of beauty in science.

Some scientific realists have argued that virtues of theories, such as simplicity, symmetry and scope, are evidential, making a theory possessing such aesthetic properties more plausible. Richard Boyd (1984) for instance. argues that theories possessing theoretical virtues that have been instantiated by previously successful theories will be ranked as more plausible. Wesley Salmon (1990) goes further in showing how these virtues can bear upon the confirmation of a theory by showing that the past record of success will ultimately determine the prior probability of a theory. Stathis Psillos (1999) endorses Boyd and Salmon's account, explaining that

> [t]he past record of mature scientific theories can be seen as the background knowledge in the light of which the plausibility of emergent scientific theories can be judged and estimated [. . .] if theories which have not been subjected to ad hoc adjustments have tended to be better supported by the evidence than theories with ad hoc features, then this consideration should be used in assessing the prior probability of other theories, in order to rank higher theories with no ad hoc features. Naturally, finding out which theoretical virtues have been associated with well-confirmed theories can be the outcome only of substantive empirical-historical research.
>
> (Psillos 1999: 172)

Independent of the scientific realism debate, James McAllister has also offered a track record account of aesthetic values. His classic work *Beauty and Revolutions in Science* (1996) offers a sophisticated account of aesthetic values in science, showing how one could justify confidence in a theory based on its beauty even if the theory has not received sufficient empirical support. The reliability of aesthetic values is supported by appealing to the established aesthetic canon, which is formed upon reflection on the aesthetic properties instantiated by the most empirically successful theories of the past. The argument relies on a mechanism known in the psychology literature as the 'exposure effect'. Studies performed on subjects have found that an agent's aesthetic preference towards an object tends to increase with repeated exposure to the object (Cutting 2003).[3] Similarly, scientists learn from their experience, by habituation and exposure, what aesthetic values have been associated with successful theories in forming an established aesthetic canon, which then guides the development and evaluation of new theories. For instance, if we reflect upon our current and past successful theories, we can place high confidence in the value of unity, since Maxwell, Newton and Einstein's theories have all achieved unification of phenomena. We can then project an expectation that when a theory comes along and has unifying power, we can expect it to be successful since it exemplifies properties conforming to the established aesthetic canon. The aesthetic canon allows us to inductively project that when a new theory exhibits the properties associated with the canon, we have good inductive ground to predict it will be empirically adequate or true even if at the time the theory lacks empirical support.

The established aesthetic canon aids inductive inferences, justifying trust in a theory that might at the time lack empirical support. By making an aesthetic induction, based on the aesthetic properties of successful theories of the past, one can infer that a theory that possesses the aesthetic properties associated with the established aesthetic canon is highly likely to enjoy long empirical success and be true. This account is both projectivist and fallibilist. It does not take aesthetic properties to have timeless and objective validity, allowing that new empirically successful theories can completely revise the accepted aesthetic canon, making aesthetic values relative to a particular framework and revisable by empirical advancement. One of McAllister's main aims is to defend the rationality of science in light of these changes in the aesthetic canon and offer an explanation of why decision making, based on aesthetic values, is rational even in light of a dynamic notion of beauty. If our aesthetic canons are revised in light of empirical findings and the improvement of our scientific theories, confidence in the aesthetic canon can at least be weakly justified inductively. McAllister takes aesthetic discontinuities to occur in theory transitions, and claims that what drives the revision of an aesthetic canon is ultimately the empirical success of the superseding theory. Theory change, and in fact scientific revolutions, are regarded as changes in the aesthetic canon, seeing theory transitions as battles between progressive scientists, who adopt a theory based on its empirical success, and conservative scientists, who delay the acceptance of a new theory because it does not conform to the established aesthetic canon. Such revolutions can result in a particular aesthetic property becoming irrelevant (such is the case of visualisation after the acceptance of quantum mechanics) or with new properties becoming part of the established canon (an example here is symmetries, which became highly desirable since the success of general relativity and quantum mechanics).

McAllister's account takes scientists' aesthetic talk literally rather than adopting a reductivist interpretation of aesthetic claims. For McAllister, we should not reinterpret the aesthetic language used by scientists to be reducible to emotive praise of the empirical success of a theory. Rather, aesthetic talk should be taken as discourse about aesthetic features of the theories. I agree with this approach and think we can motivate it on two grounds. First, in addition to the fact that scientists use aesthetic language when they praise theories, there is also evidence they undergo aesthetic experience when presented with a beautiful equation or proof.[4] Second, aesthetic considerations feature in the acceptance of theories that have not been confirmed and cannot technically be considered empirically adequate. This is particularly pressing in theoretical physics, in cases such as string theory, where there is not available evidence to support the theory and yet scientists want to place their epistemic commitment to the theory in light of its aesthetic appeal.[5] These aesthetic judgements, however, in this case cannot be claimed to reduce to empirical adequacy.

The above account offers a plausible explanation for the role of aesthetic values in science and is well motivated by the historical development of science. An important aspect of this account is that it aspired to address the question concerning the status of aesthetic judgements, whether aesthetic values are objective and fixed across different times and places, or subjective and relative to a particular individual, society or temporal framework. It has been widely recognised in philosophy of science that scientists can differ substantially with respect to how they prioritise theoretical virtues, with some valuing elegance over other virtues, while others prefer the unifying power of a theory, for instance. In such cases theory choice is contingent on the virtues most valued by the particular scientists or community, so there can be disagreement as to which theory should be preferred on the grounds of them exemplifying differently prioritised virtues. In addition, each such theory virtue can receive different interpretations in different contexts, making an objective weighting comparison between the aesthetic values of two competing theories inconclusive.[6] Defending a dynamic conception of beauty can accommodate the fact that aesthetic values seem to change during theoretical transitions. For instance, the demand for visualisation was slowly abandoned with the increasing success of quantum mechanics. Quantum mechanics has introduced a highly difficult phenomenon to visualise—the superposition of particle states. We cannot visualise a particle in a superposition, this phenomenon is counter intuitive to us, but with the theory being highly successful we can revise our aesthetic canon and no longer regard visualisability as part of this canon, allowing aesthetic appreciation of a theory that is not visualisable. Furthermore, during the Copernican revolution and Kepler's developments of the heliocentric model of the solar system, the use of perfect circles to describe the movement of the heavenly spheres was slowly replaced by ellipses, as the elliptical models became more predictively successful. This transition was not quick, since the community was reluctant to deviate from the established aesthetic canon, and endorse what was considered an imperfect geometrical shape to describe planetary motion, but such criticism slowly became irrelevant as the elliptical model gained more and more empirical support. McAllister argues that the aesthetic canon guiding the research community is ultimately formed by reflecting on the values possessed by highly successful research projects of past and contemporary science.

Leaving the aesthetic canon open to change and tying it to the empirical performance of theories seems at first sight uncontroversial. However, one can worry about the descriptive merit of this account, in particular whether it adequately describes what happens in scientific practice. The aesthetic induction, while plausible in the noted cases, seems to not be able to explain certain cases in science where some values persist and the community resiliently follows them. Such 'persistent values' are appreciated and desired in the community even if contemporary theories have failed to exemplify them. In particular, by following the aesthetic induction we would expect certain values to be abandoned or weakened, as they are not properties of contemporary successful theories. However, such a phenomenon has not always been observed in the scientific community. If aesthetic values are revisable and driven by the empirical success of the theories that exemplify them, we would expect to observe more revisions to our aesthetic canons. However, values such as simplicity and unity hold a centrality in science. Simplicity has been an ideal even before the development of successful theories and continues to be an ideal followed by scientists despite the prominence of successful theories that are not regarded elegant or simple. If the aesthetic canon is driven by the values exemplified by contemporary successful theories, we would expect the appreciation for elegance and simplicity to at least weaken and perhaps be replaced by other values. As Montano (2014) argues, certain aesthetic values have the status of 'historical constants', meaning that their appreciation in the community appears to be stable throughout theory change. This puts into question how predictively successful McAllister's account is and whether exposure to successful theories is a sufficient condition for the formation of aesthetic canons.

A common idea shared by the epistemic accounts of aesthetic values is that aesthetic values are formed by habituation. McAllister (1996), Montano (2014) and Kuipers (2002) each regard the exposure effect as the mechanism responsible for the formation of our aesthetic preferences in science. It is, however, unclear whether exposure is sufficient for the formation of such preferences in science. While exposed to successful complex theories (theories, for example, with a great number of equations or free parameters), most physicists do not report increased appreciation of complexity in their theories, on the contrary, they negatively compare complex explanations of phenomena to simple ones. In the case of the standard model, for instance, the community openly expresses the idea that the theory is aesthetically unappealing, despite its enormous empirical success. As Kaku and Thompson argue, "there has been no experimental deviation from the Standard Model. Thus, it is perhaps the most successful theory ever proposed in the history of science. However, most physicists find the Standard Model unappealing because it is exceptionally ugly and asymmetrical. [. . .] The reason why the Standard Model is so ugly is that it is obtained by gluing, by brute force, the current theories of the electromagnetic force, the weak force, and the strong force into one theory" (Kaku and Thompson 1997: 75). This dissatisfaction with the standard model is due to its failure to satisfy persistent aesthetic values, rather than what the aesthetic induction would predict, which would be that these aesthetic ideals would be revised in light of the success of the standard model.

Similarly in mathematics, as argued in Montano (2014), proofs by cases have not gained the appreciation of mathematicians, despite their successful employment. Montano focuses on case-by-case proofs to challenge McAllister's prediction that as these proofs obtain a strong record of success, we can expect them to gain aesthetic appeal in the mathematical community. Such appreciation has not been observed. What such computer-assisted proofs do is check case-by-case, which violates the parsimony criterion. In the case of Appel and Haken's proof of the four-colour theorem, the computer assists in checking hundreds of cases. The fact that the proofs violate the desideratum of simplicity, which Montano claims is associated with understanding a proof, is the reason why aesthetic appeal has not been observed despite their successful employment, challenging the prediction made by the aesthetic induction. The conclusion I draw from these two cases is that while empirical success is an important element in the formation of trust in the performance of a theory with specific aesthetic merit, it is not sufficient to explain why certain values seem persistently appreciated while others do not gain aesthetic appreciation despite being exemplified by successful theories.

Finally, my main objection to the above track-record account is methodological. The above arguments have relied on the historical record of successful theories to justify confidence in the truth of contemporary theories that exemplify certain aesthetic properties. I do not take such arguments from the history of science to be sufficient to defend an 'optimistic' over a 'pessimistic' argument. If we take the current state of the scientific realism debate as an example, we can see that arguments both in defence of and against realism have been made by inductive inferences from the history of science. Anti-realists, following Larry Laudan (1981), argue against realism on the grounds that lots of past successful theories have been abandoned, undermining the claim that success can be an indicator of truth. And scientific realists take it that no better explanation of the predictive track record of theories can be given unless we take them to be true, at least approximately. Such inductive arguments are, however, inconclusive and cannot pull towards a realist reading of science.

The problem is that there are many aesthetically pleasing theories that have turned out to be false and many aesthetically displeasing theories that fit with the highest requirements of the realist to count as true, at least 'approximately'. On the one hand, we have the beautiful theories that failed. Copernicus' heliocentric system was abandoned in favour of Kepler's despite the initial negative reaction to the idea of abandoning a system based on perfect cycles to epicycles, considered at the time aesthetically inferior. And when it comes to theoretical projects in high-level physics, unificationist projects have been abandoned due to difficulties despite being recognised for their aesthetic appeal. The Kaluza-Klein theory, for instance, provided an elegant way to unify gravity and the other gauge fields but was abandoned despite it being highly regarded due to its elegance and simplicity. Hossenfelder's (2018) *Lost in Math: How Beauty Leads Physics Astray* is full of illuminating examples from theoretical physics in which theories of great aesthetic merit are abandoned due to their lack of predictive success. Hossenfelder argues that we should be wary of relying on beauty to guide our belief in theories, and that beauty can often be a systematic bias that leads away from developing successful theories. And when it comes to placing confidence in the track record of 'ugly' theories, our current best theories in physics—the standard model and quantum mechanics—provide examples of great empirical and predictive success but are rarely used as examples of beautiful theories, quite the contrary.

The problem with drawing inductive inferences from the history of science is that we can find plenty of examples to base an inductive argument on, some in favour of the link between beauty and truth and some against it. Arguments relying on the track record of science have been well utilised in the scientific realism debate, but have not produced a conclusive choice between realism and anti-realism. Similarly in this context, it seems like inferences based on the history of science do not have the power to pull us conclusively in either direction when it comes to the question of whether there is a link between aesthetic values and truth and a further argument is needed to strengthen the realist or anti-realist position.

So far I have examined how one could support the epistemic significance of aesthetic values by inferring the likelihood of a theory from the track record of aesthetic values exemplified by previously successful theories. The objections developed above challenge the idea that there is a link between beauty and truth, but this does not necessitate accepting the idea that aesthetic values are merely heuristic. In the next section I develop an alternative account, arguing for the epistemic significance of aesthetic values. I shall argue that by reconsidering the aims of science and taking our primary aim to be understanding rather than truth, we can recognise the regulative role played by aesthetic values in achieving these epistemic aims.

4 How Beauty Guides Our Understanding

Science delivers epistemic goods, but these goods are rarely characterised as literally true theories. Often our understanding of phenomena is achieved by constructing models that deviate significantly from the truth, and the history of science teaches us that even our very best theories can turn out to be false. The fact that truth is not present in these 'layers' of scientific theorising should not be taken to undermine the epistemic success of science. This is because science has many aims, truth and empirical adequacy are certainly the usual suspects, but science also advances our understanding of the world around us. Such understanding can be achieved with theories that deviate from the truth, but this fact does not undermine their importance for the advancement of our epistemic aims. Before I defend the epistemic role of aesthetic values in science, I would like to further motivate the account of understanding I defend.

In the recent literature in philosophy of science, much focus has been given on explaining how we use models and idealisations in science. Much of the practice of making a model in science involves distortion, abstraction and idealisation, which often equates with making knowingly false assumptions. When we construct economic models, we often make assumptions such as the perfect rationality of agents. When we study the behaviour of gases, we often use laws like the ideal gas law, which represents the relationship between pressure, temperature and volume in a gas, while it falsely assumes that gases don't exhibit intermolecular attraction or that gas molecules are spherical. The use of such devices in science has led some philosophers to claim that the aim of science is not truth but rather understanding of worldly patterns, and this account can better explain scientific practice (most instructively developed in the work of Catherine Elgin (2018) and Angela Potochnik (2017).

Apart from concerns about idealisation in science, claiming that the aim of science is truth has been riddled with some well-known problems, including the problem from theory change. This problem was pressing at the beginning of the twentieth century when systematic philosophical work emerged addressing the aim of science and the limits of scientific knowledge. Reflecting on the constant revisions of theories and the apparent discontinuities occurring during such transitions, philosophers like Pierre Duhem and Henri Poincaré explicitly distanced themselves from the idea that science aims at discovering truths about that world, holding instead that scientific theories aim at prediction and lead to the deeper understanding of the relations between the phenomena.[7] Duhem and Poincaré worried as to whether we have good grounds on which to trust that our theories can reveal the true nature of reality, since many scientific theories once believed to be true descriptions of reality were abandoned and replaced by new ones. This argument, known at the time as the argument from the 'bankruptcy of science', was employed to question our optimism that scientific theories can reveal the underlying nature of reality. This problem motivated the position that the aim of science is not to offer us true theories, but to provide understanding of relations in the world.

Distancing ourselves from the traditional ideal in epistemology that sees science as obtaining truths about the world, is also motivated by the desire to make sense of the status of first principles in scientific theories. These principles are the fundamental building blocks of our theories, but are often considered to be neither true nor false. Many of the founding principles in science, the very 'first' principles, on which scientific theories are built, often receive their justification not on the grounds of experience or a priori reasoning, but by convention. The rationality principle in economics, the laws of motion in Newtonian mechanics, the light postulate in the special theory of relativity theory, and the equivalence principle in the general theory of relativity are often taken to be neither empirically true nor a priori; they lack truth value in isolation. This account of principles makes it difficult to take theories as true descriptions of reality, since the very basic principles on which these theories are built lack truth-value. There has been a lot of attention on how central principles of science are justified and what the epistemic implications are if we claim science is built on principles that are established on conventional grounds. However the status of these principles is settled, it is difficult to claim that our best scientific theories are approximately true, since the building blocks of these theories have no truth value. This problem further motivates us to reconsider the plurality of aims in science and focus our attention on epistemic goods other than truth, such as understanding.

The idea that we can have understanding without truth may appear deeply problematic at first. According to traditional epistemic accounts of understanding, truth is a necessary condition for understanding, and a kind of propositional knowledge. This account, however, is too restrictive to accommodate for the success of science. Nonfactive accounts of understanding, developed recently by Elgin (2009, 2018) and de Regt (2015), treat understanding not as a kind of knowledge but rather as a skill or ability to grasp how the facts fit together. On the face of it, there is a substantial difference between knowing a set of propositions or laws, and being able to apply these laws in different situations. This ability to 'see' how certain propositions relate to each other and apply them to different contexts is considered an essential element of understanding. Also, understanding does not always require the belief in true propositions. One's ability to use Newtonian mechanics to calculate the gravitational pull of the Moon requires an understanding of how to employ Newton's laws together with a set of initial conditions in order to derive an answer. If truth were a presupposition for having understanding of gravitational effects, one would lack understanding in this case, given that the theory in play is strictly speaking false. As Henk de Regt (2015) argues, if truth is a necessary condition for understanding, it would follow that past scientists lacked understanding of phenomena for which they had advanced empirically successful (but from our perspective false) theories. Separating the two concepts and taking understanding not to require true theories, but rather an ability to manipulate and use a theory in a certain theoretical domain, avoids this problem and leads to a more plausible thesis of the aims in science and scientific progress.

In the previous section I argued that aesthetic values cannot serve as predictors or indicators of a theory's truthlikeness. But that is not to say their use in science is not well motivated. They do in fact play a regulative role in achieving the aim of science, namely understanding the phenomena. Here I draw upon some of Poincaré's thoughts on beauty in science recently reconstructed in Ivanova (2017). According to this account, the beauty in our theories—their elegance or unity—does not reflect some elegance or harmony in the world. Rather, they are properties of our theories only because we decided to construct them in such a way, and we decide to construct them this way because it is most convenient for us to operate with theories that satisfy our aesthetic and intellectual requirements. Poincaré explicitly argues against the idea that aesthetic values can be truth indicative. Questions like 'is nature itself beautiful' are not within the empirical method to address. On the contrary, aesthetic values such as simplicity and unity should not be taken to be properties of nature itself, but conditions of our making, regulating ideas in scientific inquiry. When we construct theories that conform to the principles of unity and simplicity, we are following ideals reflective of our cognitive makeup: "[i]n formulating a general, simple, and formal law, based on a comparatively small number of not altogether consistent experiments, we have only obeyed a necessity from which the human mind cannot free itself" (Poincaré 2001: 100). He further claims that this 'harmony' we experience when we reflect on our theories "is at once a satisfaction of our aesthetic requirements, and an assistance to the mind which it supports and guides" (ibid.: 396–397). Thus, following these aesthetic values allows us to construct theories fit for achieving the goal of science, to offer us understanding of relations between the phenomena.

Taking aesthetic values in science as conditions of our cognitive makeup reflecting our intellectual interests and capacities explains why aesthetic values persist even when the best theories do not seem to quite fit our aesthetic requirements. Physicists search for a grand unifying theory because even though we have successful theories of the different forces, they are convinced our understanding of the world would be advanced by a unifying theory. Theorists in chemistry search for the best organisation of the elements even though they know all the elements, their atomic properties and structural patterns. What guides their enquiry is the idea that a more elegant representation of the elements would provide more understanding.[8] A theoretical description fitting with our aesthetic ideals seems better disposed to attain our epistemic aims.

This tendency to prefer simplicity and scope in our explanations seems deeply ingrained in human thinking and has been systematically studied in psychology and neuroscience. For instance, studies in cognitive psychology focused on an agent's preferential inferences supports the idea that agents strive to develop and consistently support simpler explanations of broader scope. Tania Lombrozo explains that "when children and adults generate and evaluate explanations, they recruit explanatory virtues, such as simplicity and breadth, as evaluative constraints on reasoning. As a result, they are more likely to generate and favor broad and simple hypotheses, and to discover broad and simple patterns" (Lombrozo 2016: 749). Our desire for simplicity and unity affects both our preferences for what theories we construct and what patterns in nature we focus on to study and identify.

This is not to say that if systematically in error, we would continue to prefer the aesthetically pleasing theory. A pretty idea that is useless would not get us far. But there is certain resistance in accepting complexity over simplicity, narrowed and more limited scope to breath. This fact is also nicely highlighted by Lombrozo's studies, arguing that agents continue to prefer simpler explanations even after they learn about the likelihood of the simple and more complex explanation and it takes a rather significant decrease of likelihood before they consider the plausibility of the complex explanation. She argues that:

> In one set of studies, participants selected between explanations for an individual's symptoms that appealed to either a single common cause (simple) or to two independent causes (complex), and they were additionally provided with information about the base rate of each disease. Under these conditions, explanation choices were sensitive to both simplicity and probability. When two candidate explanations were equally likely, a majority of participants selected the simpler explanation. It was not until the complex explanation

> was 10-fold more likely than the simpler alternative that a majority of participants selected it as the most satisfying explanation for the symptoms. It thus appears that, when the probability of an explanation is uncertain, the relative simplicity of an explanation has a significant effect on its perceived quality.
>
> (ibid.: 751)

This phenomenon can be seen in science: when aesthetically pleasing explanations systematically fail, we move on and work with theories that do not fit these requirements, but that is not to say that the search for simplicity and elegance ends.

This point can be further illustrated by considering complexity theory, where simple explanations are likely to be useless and perhaps aesthetic appreciation for complexity can be developed. Here I again draw lessons from Poincaré, whose work on the three-body problem led him to consider how complexity theory fits with the idea of simple and unified science. In the case of simplicity, for instance, he claims that apparent simplicity can conceal deep complexity in the phenomena so it cannot be considered a reliable indicator of truth: "[a] century ago it was frankly confessed and proclaimed that nature loves simplicity; but nature has proved the contrary since then on more than one occasion. We no longer confess this tendency, and we only keep of it what is indispensable, so that science may not become impossible" (Poincaré 2001: 100). Poincaré argues that even though simplicity might not be a property of nature, a question we cannot answer, it nevertheless has to be a property of our theories and we should always try to generalise in the simplest possible ways: "those who do not believe that natural laws must be simple are still often obliged to act as if they did believe it. They cannot entirely dispense with the necessity without making all generalisations, and therefore all science, impossible. It is clear that every law can be generalised in a number of ways, and it is a question of choice. The choice can only be guided by considerations of simplicity" (Poincaré 2001: 113). I thus take it that our experience of beauty, when engaging with a theory, stems from the ability of the theory to present a complex phenomenon in a simple way and give us insight into further phenomena and their relations which give us an understanding of the world.

This brings me to my conclusion. That scientists use aesthetic values in the conception and evaluation of theories is a fact. That they credit theories on non-empirical grounds is also a matter of fact. The more remote from empirical data a discipline is, the more its proponents appeal to beauty as a source of justification. These systematic preferences are quite clear in every level of scientific theorising, but it is misleading to claim that such aesthetic judgements reflect anything about the world in itself, they should be seen as reflections of our own intellectual capacities and regulative ideals guiding our epistemic aims. The quest for simplicity and unification is well grounded in our desire to understand the world and to construct theories we can easily employ and operate with. But the fact we often succeed in constructing such highly successful theories with aesthetic appeal is not a reason to infer that this beauty is a reflection of a beautiful orderly nature, as nature might be very far from this ideal.

5 Conclusion

One way to think about aesthetic values in science has been to dismiss them as subjective and irrelevant factors in the context of justification. Another has been to show not only that such values have some heuristic role to play in the context of discovery and justification, but that they can be trusted to indicate the likelihood of a theory's empirical adequacy or even truth. I have discussed the insights drawn from such empirical accounts as well as some difficulties. In this chapter I have explored a different way to address the question of whether aesthetic values can play an epistemic role in science by arguing that such values can be seen as requirements we impose upon theories for our intellectual purposes. The prominence of aesthetic values in scientists' decision-making can be seen as part of our way of thinking about the world that is deeply dependent upon our capacities as agents, independently of the question whether such values can in any way guarantee that the theories we regard as beautiful are also likely to be successful.

Acknowledgements

I am very thankful to Steven French, Alice Murphy and Matt Farr for their insightful comments on the earlier draft of this chapter. This work was presented at the universities of Bern, Cambridge, Durham, Edinburgh, Exeter and Leeds. I am grateful to the audiences for their questions and suggestions.

Notes

1. That we can find unexpected order in apparently unrelated phenomena can lead to surprise, which in this account related to the experience of beauty, offering an additional dimension to this concept and interestingly relating it to discovery and creativity.
2. While the notion of beauty in science in this quotation reflects the mathematical beauty of the physical theory, in the rest of the chapter I adopt a much more pluralistic notion of beauty, which extends to theories that are not mathematicised.
3. This point has been called into question by recent work on the exposure effect in art conducted by Meskin, Phelan, Moore and Kieran (2013). They argue that exposure to 'bad' art does not correlate with increase in subjects' aesthetic appreciation, suggesting that something over and above exposure must

be responsible for a subject's aesthetic responses to art pieces. These results are important for our understanding of beauty in science, giving us grounds to challenge the assumption that exposure and habituation are responsible or sufficient for scientists' appreciation of the aesthetic values of scientific theories.

4. Zeki, Romaya, Benincasa and Atiyah (2014) show that the same areas of the brain are as active when a mathematician is exposed to equations they have previously described as beautiful as when they are appreciating pieces of art and music.
5. Of course not all defence of the plausibility of string theory is based on its aesthetic appeal. Richard Dawid has offered a different justification for the theory, explicitly arguing against the employment of beauty as a criterion due to its subjectivity (see Dawid 2013).
6. Since Kuhn (1977), it is commonly accepted that scientists can legitimately disagree in theory choice if they favour different theoretical values (theory virtues such as simplicity, harmony, elegance, etc.). See Duhem (1954) for a discussion of theory choice and the involvement of judgement and good sense in weighting theory virtues, as well as Stump (2007) and Ivanova (2010, 2014).
7. See Ivanova (2015, 2017) and Ivanova and Paternotte (2013).
8. I am grateful to Hasok Chang and Klaus Ruthenberg for directing me to this debate.

Bibliography

Boyd, R. (1984), 'The Current Status of Scientific Realism', in J. Leplin (ed.), *Scientific Realism*. University of California Press, pp. 195–222. Berkeley: University of California Press.

Chandrasekhar, S. (1987), *Truth and Beauty: Aesthetics and Motivation in Science*. University of Chicago Press.

Cutting, J. (2003), 'Gustave Caillebotte, French Impressionism, and Mere Exposure', *Psychonomic Bulletin & Review*, 10: 319–343.

Dawid, R. (2013), *String Theory and the Scientific Method*. Cambridge University Press.

De Regt, H. (2015), 'Scientific Understanding: Truth or Dare?', *Synthese*, 192: 3781–3797.

Dirac, P. A. M. (1980), 'The Excellence of Einstein's Theory of Gravitation', in M. Goldsmith, A. Mackay and J. Woudhuysen (eds.), *Einstein: The First Hundred Years*. Oxford: Pergamon Press.

Duhem, P. (1954 [1906]), *The Aim and Structure of Physical Theory*. Princeton University Press.

Elgin, C. (2009), 'Is Understanding Factive?', in D. Prichard, A. Millar and A. Haddock (eds.), *Epistemic Value*. Oxford University Press.

Elgin, C. (2018), *True Enough*. MIT Press.

Heisenberg, W. (1971), *Physics and Beyond: Encounters and Conversations*, trans. Arnold J. Pomerans. George Allen and Unwin.

Hossenfelder, S. (2018), *Lost in Math: How Beauty Lead Physicists Astray*. Basic Books.

Ivanova, M. (2010), 'Pierre Duhem's Good Sense as a Guide to Theory Choice', *Studies in the History and Philosophy of Science*, 41: 58–64.

Ivanova, M. (2014), 'Is There a Place for Epistemic Virtues in Theory Choice?', in A. Fairweather (ed.), *Virtue Epistemology Naturalized*. Synthese Library, Vol. 366, pp. 207–226.

Ivanova, M. (2015), 'Conventionalism About What? Where Duhem and Poincaré Part Ways', *Studies in the History and Philosophy of Science*, 54: 80–89.

Ivanova, M. (2017), 'Poincaré's Aesthetics of Science', *Synthese*, 194: 2581–2594.

Ivanova, M. and Paternotte, C. (2013), 'Theory Choice, Good Sense and Social Consensus', *Erkenntnis*, 78: 1109–1132.

Kaku, M. and Thompson, J. T. (1997), *Beyond Einstein: The Cosmic Quest for the Theory of the Universe*. Oxford University Press.

Kuhn, T. (1977), 'Objectivity, Value Judgment, and Theory Choice', in *The Essential Tension*. University of Chicago Press, pp. 320–353.

Kuipers, T. (2002), 'Beauty, a Road to the Truth', *Synthese*, 131: 291–328.

Laudan, L. (1981), 'A Confutation of Convergent Realism', *Philosophy of Science*, 48: 19–49.

Lombrozo, T. (2016), 'Explanatory Preferences Shape Learning and Inference', *Trends in Cognitive Science*, 20: 748–759.

Mach, E. (1984), *The Analysis of Sensations and the Relation of the Physical to the Psychical*, trans. C. M. Williams. La Salle: Open Court.

McAllister, J. (1989), 'Truth and Beauty in Scientific Reasoning', *Synthese*, 78: 25–51.

McAllister, J. (1996), *Beauty and Revolution in Science*. Cornell University Press.

Meskin, A., Phelan, M., Moore, M. and Kieran, M. (2013), 'Mere Exposure to Bad Art', *British Journal of Aesthetics*, 53: 139–164.

Montano, U. (2014), *Explaining Beauty in Mathematics: An Aesthetic Theory of Mathematics*. Synthese Library, Vol. 370.

Poincaré, H. (2001), *The Value of Science: Essential Writings of Henri Poincaré*, ed. Stephen Gould. New York: Modern Library.

Potochnik, A. (2017), *Idealization and the Aims of Science*. University of Chicago Press.

Psillos, S. (1999), *Scientific Realism: How Science Tracks Truth*. Routledge.

Salmon, W. (1990), 'The Appraisal of Theories: Kuhn Meets Bayes', *PSA: Proceedings of the Biennial Meeting of the Philosophy of Science Association*, 2: 325–332.

Stump, D. (2007), 'Pierre Duhem's Virtue Epistemology', *Studies in History and Philosophy of Science*, 38: 149–159.

Zeki, S., Romaya, J. P., Benincasa, D. M. T. and Atiyah, M. F. (2014), 'The Experience of Mathamtical Beauty and Its Neural Correlates', *Frontiers in Human Neuroscience*, 8: 1–12.

6 A Plea for the Sublime in Science

Margherita Arcangeli and Jérôme Dokic

1 Introduction

There is a growing literature on the role of aesthetic values, experiences and judgements in the scientific endeavour. The aesthetic dimension of science is suggested by scientists themselves. Indeed, they often describe their objects of study or scientific achievements in terms of beauty, using related adjectives like "harmonious", "simple" or "wonderful". Sometimes they even go further and make direct parallelism between artworks and their theories, laws and experiments. Scientists have also praised their aesthetic sensibility, referring in their decision-making process to emotions and feelings involved in aesthetic experiences (e.g., pleasure, delight, contemplation, exaltation, wonder, awe). Moreover, often enough scientists hold that such aesthetic dimension reveals a connection with the epistemological (and potentially even the ontological) dimension of science: there would be an intimate relationship between beauty and truth.

Aesthetics seems to enter science on at least three different levels:

(i) The *objects* of scientific enquiry (such as cells, mu-mesons and numbers) may instantiate aesthetic values.
(ii) The *products* of science (such as theories, conjectures and models) may instantiate aesthetic values.
(iii) The scientific *practice* (such as constructing and evaluating theories, and designing experiments) may be guided by aesthetic experiences and judgements.

There is arguably a tight connection between the first two levels, not least because the objects of scientific enquiry are often *theoretical* entities. For instance, a theory might inherit aesthetic properties, such as simplicity, order and coherence, from the fragment of reality that it concerns. The theory would be beautiful because it describes something beautiful. The first two levels belong firmly to the aesthetic domain, and they might provide some kind of justification to the third level. If either the objects or the products of science bear aesthetic values, then aesthetic experiences and judgements about them may be apt to guide the scientific practice, whether in the construction or in the evaluation phase.

Of course, without a clear picture of what aesthetic values, experiences and judgements are, it is hard to assess the real import of aesthetics for science at all these levels. The aim of this chapter is to lay a bridge between recent discussions within naturalised aesthetics (i.e., aesthetics with an eye to producing empirically testable hypotheses) and the debate in philosophy of science on the aesthetic dimension of science.

More precisely, we argue that both literatures have focused almost exclusively on the beautiful and have neglected another important aesthetic category, namely the sublime. The latter can be legitimately considered as belonging to the aesthetic domain, which arguably is variegated and not exhausted by the beautiful. We will show how endorsing such an aesthetic pluralism enriches the debate in philosophy of science.[1]

The structure of our discussion is as follows. In section 2, we introduce the philosophical distinction between the beautiful and the sublime as distinct aesthetic types, which correspond to aesthetic experiences with different cognitive profiles. In section 3, we address two questions: whether science deals with objects qua bearers of aesthetic values, and whether scientists undergo aesthetic (i.e., both beauty and sublimity) experiences in their endeavour. The discussion will lead to consider that both beauty and sublimity aesthetic experiences have become scientific objects of empirical aesthetics. In section 4, we turn to the issue of *how* aesthetic experiences and judgements can guide the scientist in her endeavour, tackling the influential idea that there is an intimate link between beauty and truth, as well as the emerging idea that aesthetic judgements are rather connected to scientific understanding.[2] In so doing we shall discuss an influential empirical model of aesthetic experience, which hinges on the notion of processing fluency, and point to its limits in dealing with sublimity experiences. Our claim is that the relational nature of such experiences is such that they can play a specific role, in addition to experiences of beauty, in the construction and evaluation of theories and other scientific products.

2 The Beautiful and the Sublime

A merganser diving in a clear lake, shining patterns made by light reflections on a wall, a colourful countryside landscape, can elicit pleasurable aesthetic experiences. It proves difficult to say why these experiences are aesthetic. Naturalised aestheticians have extensively drawn on empirical research to inform philosophical theories of aesthetic experiences and have put forward very different accounts (for a recent illustration, see Cova and Réhault 2018). Despite their diverging views, naturalised aestheticians tend to agree in thinking that aesthetic experiences, in contrast to non-aesthetic experiences, are characterised by an attentional, self-sustaining or "auto-telic" pattern (see Schaeffer 2015 and Dokic 2016, a similar point is also made in Prinz 2011). The idea is that when we undergo an aesthetic experience, we are motivated to maintain the relation with the object that triggered such an experience, and our motivation is internal to the experience itself. As Kant put it, we linger in the contemplation of the beautiful. This seems precisely what happens in the aforementioned examples.

Contrast the aesthetic experience we can have in admiring the merganser movements or the light reflections with the experience we can undergo when facing the stars in a pure night sky, huge and steep mountains, the primeval force of waterfalls, or the majesty of a T-rex skeleton. In these circumstances, our experience seems to be tinged with negative feelings, and fear and admiration seem to mix together. There is an overwhelming vastness, or power, which disturbs and unsettles our mind. In his *Lectures on Physics*, Richard Feynman describes the experience of contemplating the stars in a desert night as follows: "The vastness of the heavens stretches my imagination—stuck on this carousel my little eye can catch one-million-year-old light. A vast pattern—of which I am a part" (Feynman et al. Ch. 3, §4, fn. 1). We are reminded that we are just human beings confronted with, as Stephen J. Gould would put it, "a majestic entity of such vast spatial and temporal scope that she cannot care much for a little mammalian afterthought with a curious evolutionary invention" (Gould 1991: 13). Moreover, despite their negative aspect, these experiences are also the kind of experience that we want to sustain and seek out, that is, they show the auto-telic character proper to aesthetic experiences, thus suggesting that they are indeed aesthetic experiences.

The contrast between these two types of aesthetic experience reveals the variety of aesthetic relationships we can have with the world. Philosophers have captured such a contrast with the distinction between beauty experiences and sublimity experiences (for a recent historical account of this contrast, which is rooted in ancient rhetoric, see Brady 2013). Beauty experiences concern things mentioned in the first examples we gave (a merganser diving, light reflections on a wall, a colourful countryside landscape). Features which have been put forward in order to characterise the objects of beauty experiences include delicacy, smoothness, proportion, fragility, harmony (for discussion, see Scruton 2009 and Levinson 2011). In general, objects of beauty experiences seem to be smaller in size, scope, or power than the objects of sublimity experiences. Thus, beauty experiences lack the overwhelming aspect present in sublimity experiences and, in turn, do not show the negative aspect shown by the latter.[3] There is nothing disturbing or unsettling when we are struck by the beauty of a merganser diving. Beauty experiences are mainly positive and pleasurable, are delighting and invigorating, and involve reward and satisfaction. This seems to be true even for cases of so-called "terrible beauty" (see Brady 2013: Ch. 7), that is beauty experiences triggered by more disturbing objects (e.g., a colourful pattern rising on a muddy, polluted pond).

The core elicitor of sublimity experiences seems to be vastness. In all the examples of the second type that we offered, there is a confrontation with something that overwhelms us, in size (as the mountains), scope (as the stars) or power (as the waterfalls). Although natural scenes are considered as paradigmatic elicitors of sublimity experiences, arguably also human creations can trigger them. This would be the case, when, for instance, we contemplate the greatness of the Great Wall of China, we stand underneath (or on top of) the Eiffel Tower, or even when we grasp the deep meaning of a scientific theory (e.g., General Relativity).

We have already stressed that the *grandeur* we face in sublimity experiences has a negative effect, given its disturbing and unsettling aspect. Very often philosophers have talked about fear in order to grasp the negative feeling involved in sublimity experiences. The idea might be that it is as if we could foresee a potential danger, though (we judge that) there is no real danger. For instance, the wrath of the waterfalls might turn against us, or we might slide down the steep mountains. However, reference to fear does not seem to capture all examples (see Cochrane 2012). In some cases, we might just feel uneasiness in being confronted with something greater, in time or space, than us (e.g., as in the stars, the T-Rex or the Great Wall of China cases). As suggested by both Feynman's and Gould's remarks, a sense of one's smallness may arise and we may even feel the insignificance of human life. Similarly, in the General Relativity case we may be bewildered by the overturning of our ordinary way to think of time and space, which makes us uncertain about our place in the universe. In addition to the feeling of losing one's grip on these most basic elements of reality, we might experience our smallness or insignificance relative to the genius who achieved such a conceptual revolution. Our confrontation with a theory manifestly designed by a mind so much greater than ours may have the disturbing effect on ourselves that is characteristic of sublimity experiences.

However, the greatness we are confronted with has a positive effect too. It raises a challenge to our mind, which is enlightening and elating. Our senses, intellect and imagination have to "stretch" (to echo Feynman's quote) themselves to handle, for instance, immense expanses, a myriad of objects, overpowering forces and astonishing achievements. Shortly after the given quote, Gould says: "Thus, I love nature primarily for the puzzles and intellectual delights that she offers to the first organ capable of such curious contemplation" (Gould 1991: 13). Similarly, Joseph Addison wrote: "Our Imagination loves to be filled with an Object, or to grasp at any thing that is too big for its Capacity" (1712). It has been underlined that sublimity experiences seem to disclose new levels of knowledge, or even reality. Supernatural values, as Kantian philosophy suggests, or the essence of the universe and the natural forces that govern it, as it emerges in Schopenhauer's view (Schopenhauer 1819/1844), might be revealed by sublimity experiences. The German term for "sublime" is precisely "Erhabene", which is etymologically tied to the noun "Erhebung", meaning elevation. It is not surprising that the experience of the sublime has been associated with spiritual or mystical experiences.

The positive aspect of sublimity experiences shows that they bear some similarities with beauty experiences. Notwithstanding the double aspect (positive and negative) of the sublime, arguably its overall valence is positive. Perhaps it is not accurate to talk about pleasure to capture such a positive valence—Kant stressed that the sublime seems to involve a kind of pleasure "that is only possible by means of a displeasure" (Kant 1790: §27). As stressed beforehand, it is obviously not the kind of experience that we want to cease to have. On the contrary, we want to keep it alive as long as possible or to reproduce it, as in the case of beauty experiences.

We have ended up with a rough idea of what an aesthetic experience is, and of its two varieties, corresponding to different cognitive patterns (involving sensory, attentional, emotional and intellectual aspects), namely beauty and sublimity experiences. Let us also assume that these two patterns correspond to different aesthetic properties or values (as our examples below will make clear). In the following sections we will look more closely at the impact of this picture on the aesthetics of science.

3 Aesthetic Objects of Scientific Enquiry

The speed of light, the structure of DNA, the nature of gravity, mammalian evolution, mathematical constants, etc. are all objects of scientific enquiry diverging in scale and scope, but can we find a common denominator? More precisely, given our focus in the present context on the aesthetic dimension of science, could we say that science deals with objects qua bearers of aesthetic values? The idea might even be pushed further, and it might be suggested that scientific objects show aesthetic values and these very values capture the scientist's attention leading her to study those objects. Scientists' reports of their practice hint at such an idea. Aesthetic values such as harmony, regularity, coherence, unity and simplicity are often invoked by scientists as possessed by the objects under scrutiny and driving their enquiry. (As we have observed in the introduction, such values can also be instantiated, perhaps by inheritance, by the scientific constructions themselves, namely theories, hypotheses, models, etc.).

This is a vexed issue. Strong sceptics claim that when talking about aesthetic values scientists are not really referring to properties they are acquainted with, but rather merely employing metaphorical language. It would be a sort of confabulation, which hides other values or reasons (e.g., pragmatic or epistemic) behind aesthetic language. Leaving aside such a view and granting that scientific objects have aesthetic values, there is another pressing worry concerning what aesthetic values are relevant in the scientific domain. They might depend on the specific discipline. Physicists and mathematicians often praise aesthetic values like regularity and simplicity, but irregularity and complexity can also be seen as important aesthetic values shown by scientific objects. This is clearly seen if we turn, for instance, to biology (Ivanova 2017b), but even to specific branches of physics itself (e.g., Quantum Theory). Moreover, even within the same scientific discipline different levels of enquiry can show different aesthetic values. For instance, electricity might be seen as bearing different aesthetic values in different contexts (e.g., when conducting an experiment and when developing a theory).

To add a further layer of complexity, aside from the issue of scientific objects qua bearers of aesthetic values, we might ask whether in developing theories or studying phenomena scientists undergo aesthetic experiences. Again, scientists' reports point in that direction. Albert Einstein, for instance, said that: "The fairest thing we can experience is the mysterious. It is the fundamental emotion that stands at the cradle of true art and true science" (Einstein 1934/1935: 5—see also Einstein 1932). Although Einstein is not explicitly talking about an aesthetic experience as such, he is referring to a kind of experience, which, in his view, is crucially involved in both art and science. Other scientists have compared the experience triggered by the confrontation with art to the experience triggered by the confrontation with science. To give an example, "Mathematics, rightly viewed, possesses not only truth, but supreme beauty", wrote Bertrand Russell (Russell 1919: 60). The aside "rightly viewed" suggests that what matters is the perspective, the experience that the subject is undergoing in being confronted with mathematics. The hypothesis that the aesthetic experiences mathematicians and artists have described with similar words are very close, if not the same, has been recently investigated by neuroscientists.

Aesthetic experiences themselves can be the object of scientific scrutiny. A new trend in aesthetics is empirical aesthetics, which attempts to understand aesthetic experiences at the neurological level. An experimental study conducted by Semir Zeki, Romaya, Benincasa and Atiyah (2014) directly investigates the neural correlates of mathematical beauty. Drawing on previous data on the neural correlates of beauty experiences elicited by visual and musical stimuli and recording neural activity by means of functional Magnetic Resonance Imaging (fMRI), Zeki and colleagues show that the same brain area is activated when mathematicians are presented with mathematical formulae.[4] These findings offer important insights into the nature of aesthetic experiences and (assuming that we can generalise to other sciences) lend force to the idea that scientists' talking in aesthetic terms should be taken literally as referring to aesthetic experiences.

Both the debate over aesthetic values in science and the debate over aesthetic experiences in scientific practice focus exclusively on beauty experiences and the aesthetic values that are typically tied to them, such as coherence, unity and simplicity. However, the topic of the sublime and the aesthetic pluralism we are endorsing sheds new light on both issues.

First, we have seen that, although scientists praise aesthetic values like regularity and simplicity more often than values like irregularity and complexity, the question is not settled, since the latter can also matter aesthetically. Moreover, note that such a tension between aesthetic values pulling in opposite directions raises a genuine concern only if we think that they belong to the same aesthetic type. However, it might be claimed that while values like regularity and simplicity are tied to the beautiful, values like irregularity and complexity are more characteristic of the sublime, given its disturbing, unsettling and challenging aspects. Therefore, acknowledging that the aesthetic domain encompasses more than the beautiful opens up new paths of inquiry in the domain of aesthetic values involved in science.

Second, it also seems promising to give room in science for sublimity experiences as genuine aesthetic experiences along with beauty experiences. It is striking to notice that scientists themselves use words evocative of sublimity experiences when they describe their scientific experiences. Einstein (1932), for instance, talks about the "mysterious" as "the most beautiful experience" grounding "all serious endeavour in art and science". But arguably the mysterious can be tied to sublimity experiences, given that they involve a confrontation with a greatness, which overwhelms and bewilders us, in a way that challenges our mind and eventually enlightens us. Einstein goes on and describes the mysterious as follows: "To sense that behind anything that can be experienced there is a something that our mind cannot grasp and whose beauty and sublimity reaches us only indirectly and as a feeble reflection". Here the fact that in sublimity experiences our mind is prompted to cope with a grandeur beyond its own power, the overwhelming aspect of sublimity experiences, is explicitly stressed. Moreover, Einstein thinks that the mysterious also underlies religion, and as we have observed, sublimity experiences have been associated with spiritual experiences.

Other telling words come from Richard Feynman, when he comments on James Watson's report of what he experienced during the DNA structure discovery. He writes:

> Is the sudden transformation of all the relevant scientific characters from petty people to great and selfless men because they see together a beautiful corner of nature unveiled and forget themselves in the presence of the wonder? (. . .) But when you describe what went on in *your* head as the truth haltingly staggers upon you and passes on,

> finally fully recognized, you are describing how science *is* done. I know, for I have had the same *beautiful* and *frightening* experience.
>
> (Feynman 2005, to James D. Watson, February 10, 1967—the latter two italics are ours)

In this passage we retrieve the double, positive and negative, nature of sublimity experiences. Feynman also hints at wonder, which is another emotion frequently associated with the sublime, and with aesthetic experiences in general (Prinz 2011).[5]

Similar words can be found in Michael Faraday's description of his emotional state when engaged with experiments on Gravity:

> I have been arranging certain experiments in reference to the notion that Gravity itself may be practically and directly related by experiment to the other powers of matter and this morning proceeded to make them. It was almost with a *feeling of awe* that I went to work, for if the hope should prove well founded, how great and mighty and sublime in its hitherto unchangeable character is the force I am trying to deal with, and how large may be the new domain of knowledge that may be opened up to the mind of man.
>
> (Faraday V, 156—our italics)

Faraday here is pointing at the enlightening and elating aspect of sublimity experiences, due to the challenge our mind is called to face. The reference to the feeling of awe, however, suggests that such a positive experience is tinged with a negative feeling. It is difficult to define what awe is. Interestingly, in some languages other than English there is no one-word translation of awe and an expression conveys its positive and negative components. For instance, in French "awe" can be rendered by a complex phrase meaning something like "fear mixed with admiration" ("effroi mêlé d'admiration"). Psychologists have mentioned awe as the specific emotion triggered by the sublime. For instance, in their comprehensive review of studies about awe in different theoretical domains (e.g., psychology, philosophy, religion and sociology), Dacher Keltner and Jonathan Haidt (2003) explicitly mention the connection between the philosophical concept of the sublime and awe, and propose as prototypical aspects of the latter vastness and need for accommodation, which echo power and obscurity (in the metaphorical sense of being difficult to grasp by intellect) in Burke's definition of the sublime.

Keltner and Haidt's work shows that sublimity experiences themselves can be the object of scientific enquiry. This is extremely important, because only by getting clear on the nature of sublimity experiences would we be in the position to clarify whether scientists are indeed reporting a sublimity experience when using sublime-related vocabulary. In such a way we would keep at bay sceptical stances according to which such a vocabulary is either merely metaphorical or empty (Todd 2008).

Empirical aesthetics seems to be well placed to offer important insights into the mechanisms that allow us to undergo sublimity experiences. However, although there is a growing body of work in this field, this approach seems to have conflated aesthetic experiences with beauty experiences, since it has produced abundant data mostly on the neural correlates of the latter. A notable exception is an experimental study conducted by Tomohiro Ishizu and Semir Zeki (2014), which directly investigates the neural correlates of the sublimity experience. In line with the philosophical picture, Ishizu and Zeki found: the activation of brain areas corroborating the double (positive and negative) nature of sublimity experiences, a weak correlation with pleasantness, but a strong correlation with grand scale. However, their findings diverge from the philosophical literature on two points. First, in their study Ishizu and Zeki have found that sublimity experiences do not activate the areas active during beauty experiences. For this reason, they take sublimity experiences to be non-aesthetic. Second, their neuroscientific findings suggest that experience of the sublime involves suppressed or at least diminished self-reflection. This is in contrast with the philosophical idea that this experience is self-centred, being frequently associated with the feeling of the insignificance of human life, of our own smallness compared to the *grandeur* we are confronted with.[6]

Both points are not as problematic as they may seem at first sight. First, though Ishizu and Zeki's neuroimaging findings did not show the recruitment of overlapping brain structures between sublimity and beauty experiences, this comparison was made on the basis of results concerning two different studies employing quite different material. In the study on sublimity, stimuli consisted in (pictures of) natural scenes, while in the study of beauty, (pictures of) paintings were employed. Although of course more philosophical work is needed to characterise more precisely what aesthetic experiences are, their results are fully compatible with the ontological view that sublimity and beauty experiences are two species of the same genus, viz. aesthetic experience. Thus, to date, no firm conclusion could be made on this issue.

Second, an account can be offered of how the self is involved in the sublimity experience that accommodates both the empirical observation that such experience involves decreased self-focused attention (recall Feynman's words: scientific endeavour makes scientists forget "themselves in the presence of the wonder") and the claim that they are self-centred (giving rise to the sense of one's own insignificance or smallness). Arguably while beauty experiences are rather object-centred, sublimity experiences seem to be much more self-centred. The latter seem to result from irreducibly relational properties involving the subject's self and her immediate environment (e.g., not just the *Garganta del Diablo*, but these waterfalls in comparison with one's smallness).[7] In contrast, the features that are responsible for beauty experiences are (at least mostly) in the beautiful things themselves (e.g., harmony, simplicity and symmetry are perceived as objective rather than subjective features). However, this does not mean that when we undergo a sublimity experience we are aware of the self-relative properties that trigger such an experience. When the subject feels the majesty and power of Iguaçu Falls, she tends to consider the waterfalls themselves as the object of her experience, without any apparent contribution of the self, while in fact it concerns herself in relation to the immediate environment.

Putting things this way echoes what Kant said about sublimity experiences. He claimed that sublimity "is not contained in anything in nature, but only in our mind" (Kant 1790: §28), thus suggesting that this type of experience appears to be object-centred, but in fact concerns only the human mind. Against a radical take on experiences of the sublime, which puts too much emphasis on the subject side, we have proposed (see Arcangeli, Dokic and Sperduti 2018) a less radical account without falling prey to the opposite radicalism, which pays too much attention to the object side and underestimates the subject side.[8] Sublimity experiences can be self-centred at the *ontological level* (i.e., their objects are relational properties essentially involving the self), but object-centred at the *phenomenological level* (i.e., subjects "forget themselves" and attribute sublimity exclusively to the environment).

On this view, the decreased self-focused attention found by Ishizu and Zeki can be explained by the fact that sublimity experiences, contrary to beauty experiences, have implications for the status of the self, insofar as they are ontologically more self-centred than beauty experiences. The pressing question is why phenomenologically the self seems to disappear. Our tentative hypothesis is that we have decreased self-focused attention, because sublimity experiences are immersive, they tend to blur the phenomenological boundary between the self and the world. The sublime overwhelms us, to the point that we lose ourselves in it.

All these considerations highlight the importance of the aesthetic category of the sublime, and also lay the basis for fruitful interdisciplinary research, which is very much needed to get clear on the nature of the sublime and on its role in science. As stressed beforehand Ishizu and Zeki's study is the first experimental study on the sublimity experience in the psychological domain. It only considers paradigmatic elicitors of such an experience, namely natural scenes, without taking into account other elicitors, such as artworks or scientific objects. Thus, the psychological literature neglects the role that sublimity experiences might play in scientific practice. In the following section we would like to focus on how aesthetic experiences can intervene in scientific activities with an eye to both beauty and sublimity experiences.

4 Aesthetic Experiences in Scientific Practice

4.1 The Fluency-Based Account in Psychology

Why should aesthetic considerations be relevant to the scientific practice? Traditional answers to this question point to ontological connections between aesthetic and epistemic values. Consider the venerable Platonic view that beauty and truth are the same. On this view, theories (hypotheses, conjectures, etc.) endowed with aesthetic properties, such as simplicity, elegance, symmetry, etc., would also be true. A view which is slightly less strong has it that beauty at least partly constitutes truth: theories that lack aesthetic properties cannot be true.

Interesting epistemic *norms* can be grounded on such ontological connections. If truth is or entails beauty, then, as Feynman puts it, "you can recognize truth by its beauty and simplicity" (quoted in Schwarz 2018). On this view, evidence that a theory is beautiful is also at least partial evidence that it is true. Specific aesthetic properties can guide the scientists to the truth of their theoretical constructions, either when they are developing or evaluating them. For instance, faced with extensionally equivalent but internally different theories, the scientist *should* believe, or would be *justified* in believing, the more simple or elegant one. More generally, we are at least *prima facie* justified in accepting (or rejecting) a theory because of its positive (or negative) aesthetic properties.

Independently of the existence of such epistemic norms, a psychological question arises as to whether scientists actually use psychological *heuristics* or "rules of thumb" connecting the pursuit of truth with aesthetic experiences and judgements. For instance, a theory that is assessed as beautiful, or more beautiful than another theory, will also be judged to be more truthful or faithful to the facts. The scientist's psychological assessment can be implicit or explicit. In the former case, she is biased toward beautiful theoretical constructions but is not necessarily aware of this bias. In the latter case, her assessment will typically take the form of an aesthetic judgement: a theory that is *judged* to be beautiful will also be judged to be true.

Note that the psychological question does not entail the existence of genuine epistemic norms grounded on ontological connections between beauty and truth. Even if aesthetic values do not exist or do not give rise to genuine epistemic norms, it might be that, as a matter of fact, scientists use aesthetic assessments to build and evaluate theories, hypotheses, etc. Richard Feynman's claim can of course be interpreted in the epistemological sense (beauty is an epistemic norm because it co-varies with truth), but it might also be the expression of a mere psychological heuristic: "Believe only what is beautiful (ceteris paribus)".

The psychological interpretation is empirically testable. Consider the fluency-based account of judgements of beauty and truth (for a recent statement, see Schwarz 2018). The core idea of this account is that "judgements of beauty and judgements of truth share a common characteristic: people make them, in part, by attending to the dynamics of their own information processing" (p. 25). The relevant dynamic features have to do with so-called "processing fluency". Some psychological processes are more fluent than others. For instance, it is easier for the visual brain to process the face of a friend than that of a stranger. Likewise, a long and complex sentence will be difficult to process in contrast to a short and simple one. Processing fluency is something that can transpire at the conscious level. There is more to the phenomenology of perceiving or thinking than what is perceived or thought. More precisely, processing fluency can be *felt*. Our friend's face feels familiar, unlike the stranger's. A long and complex sentence feels difficult to parse and to understand, in contrast to a short and simple one, etc.

Now there is some empirical evidence that processing fluency is the source of at least some judgements of beauty (Reber 2012; Bullot and Reber 2013). For instance, a visible shape will be judged more beautiful than, or aesthetically preferable to, another visible shape if the former is easier to process than the latter. Ease of processing may be due to intrinsic features of an object, such as its simplicity, symmetry, balance, clarity, contrast, etc., but also to whether the subject is used to process the object. According to the so-called "mere exposure effect" (Zajonc 1968), the more we perceive an object, the more we like it: mere exposure to the object can make it appear more beautiful, or at least aesthetically agreeable.

Interestingly, it seems that processing fluency is also the source of at least some judgements of truth. When a sentence is easy to process, the subject will tend to judge it as true (absent other information on the subject-matter). Moreover, the mere exposure effect can work for judgements of truth as well. Ceteris paribus, the more we are confronted with a rumour, the more we tend to accept it as true (Allport and Lepkin 1945, cited in Schwarz 2018). According to the fluency-based account, any parameter that increases the ease with which an object is processed should also increase the likelihood that the object is experienced as beautiful or accepted as true.

Let us assume, then, that some judgements of truth and some judgements of beauty are psychologically rooted on the same type of feeling, having to do with the dynamics of our own information processing. Does it follow that aesthetic experiences and judgements guide the scientific practice? The implication is not straightforward. Note that the way the feeling of fluency is spontaneously interpreted by the subject is a matter of contextual variation. Depending on the context, such feeling can evoke either mere ease of processing, familiarity, truth or beauty. The scientist who is building or evaluating a theory might just manifest a general preference for fluent processing without having any aesthetic experience as such. In other words, feelings of fluency might guide the scientific practice independently of whether they are interpreted as aesthetic experiences and feed spontaneous aesthetic judgements. Moreover, it is not clear that a particular feeling of fluency can be *simultaneously* interpreted as a feeling of truth and as a feeling of beauty. In general, feelings of fluency get an aesthetic interpretation, and feed spontaneous judgements of beauty, only if the subject is *ignorant* about the cause of fluency (Reber 2012). For instance, the mere exposure effect works only if the subject does not know that processing fluency has been enhanced by mere repetition. When the subject is aware of the cause of fluency, her feeling of fluency is interpreted as simple familiarity with the object. Thus, what counts as an appropriate context for a feeling of fluency to feed a spontaneous aesthetic judgement might not count as appropriate for it to feed a spontaneous judgement of truth.

Still, we can argue that a two-step mechanism can connect the fluency-based account with the claim that aesthetic experiences and judgements guide the scientific practice. The first step of the mechanism is the subject's implicit realisation, across several contexts, that the same type of feeling underlies both judgements of beauty and judgement of truth. The second step is the interiorisation of a heuristic connecting beauty and truth, something like "If it feels beautiful, it must be true". The point of the heuristic is to *extend* one's ability to form judgements of truth. Judgements of truth can be based on feelings of truth (i.e., on feelings which in the relevant context are interpreted as truth-conducive), but thanks to the heuristic, they can also be based on feelings of beauty. The fluency-based account explains the first step of the mechanism, and thus contributes in part to the explanation of the more general claim that aesthetic experiences and judgements guide scientific practice.

As it stands, the heuristic "If it feels beautiful, it must be true" is a mere feeling-based psychological shortcut. Can it be elevated to a general epistemic norm connecting truth and beauty? The answer depends on whether processing fluency can be said to track both truth and beauty. As we have observed, processing fluency is a function of objective properties, such as simplicity, symmetry, order, etc., but it is also the function of subjective properties, such as whether and to what extent the subject has processed the object, or similar objects, in the past. The hard question is whether such idiosyncrasy is compatible with the claim that processing fluency is a reliable sign of both truth and beauty.

The fluency-based account points to a common psychological core of judgements of beauty and truth, but as many authors have suggested (including Ivanova 2017b and Breitenbach 2013), aesthetic considerations may aid scientific *understanding* rather than truth. The ordinary notion of understanding is truth-independent: for instance, one may understand a sentence (such as "There is life on Alpha Centauri") even if we do not know whether it is true or false. However, there might be a deeper notion of understanding as involving "an ability to grasp how the facts fit together" (Ivanova 2017b: 6). On this notion, understanding a theory is more than just understanding the syntax and semantics of each sentence constituting it. Rather, it demands that we grasp the theory as a unified relational whole beyond the separate contributions of its components. It is controversial how the deeper notion of understanding relates to truth and to knowledge. One view is that understanding in this sense *entails* truth. A less radical view is that it is at least *truth-conducing*: deep understanding (from now on, just "understanding") would thus be a symptom of truth. Independently of this controversy, the fluency-based account can easily be extended to judgements of understanding (or at least some of them; see below). The notion of understanding has been related to values such as coherence, unity and simplicity, which are correlated with processing fluency at the psychological level. It is perhaps not an accident, then, that some judgements of understanding, whether or not they also involve judgements of truth, can have common roots in processing fluency as spontaneous aesthetic judgements of coherence, unity and simplicity.

4.2 *The Heuristic Role of Sublime: Disfluency*

The fluency-based account is about judgements of beauty, but what about the sublime? It is widely agreed that our experience of the sublime crucially hinges on negative emotions, either fear or terror, or a more general affective experience such as a "feeling of self-negation" (Cochrane 2012). It seems to follow that sublimity experiences are associated with *disfluent* rather than fluent processing, which nonetheless eventually leads to aesthetic pleasure. Recall Burke's famous contrast between beauty and sublimity:

> For sublime objects are vast in their dimensions, beautiful ones comparatively small; beauty should be smooth, and polished; the great, rugged and negligent; beauty should shun the right line, yet deviate from it insensibly; the great in many cases loves the right line, and when it deviates, it often makes a strong deviation; beauty should not be obscure; the great ought to be dark and gloomy; beauty should be light and delicate; the great ought to be solid, and even massive.
>
> (Burke 1759: 113)

It is clear from this passage that beauty experiences are on the side of fluency whereas sublimity experiences are on the side of disfluency. Typically at least, small, smooth, polished, light and delicate objects are processed fluently, but rugged, negligent, dark, gloomy and powerful objects are processed disfluently.

Defenders of the fluency-based account can envisage at least two strategies in order to deal with the sublime as an aesthetic category. The first strategy is *restrictive* and insists that the domain of the account is limited to a specific category of aesthetic mental phenomena, namely experiences and judgements of beauty. On this view, sublimity experiences and judgements should be the topic of another, quite different account. Any such account would have to accommodate the apparent asymmetry between fluency and disfluency regarding their aesthetic interpretation. On the fluency-based account, fluency intrinsically feels good, and thus is apt to feed spontaneous aesthetic judgements (of beauty). In contrast, disfluency does not feel so good (it is generally a signal that something is novel, or wrong), so that it is much less clear how they could feed spontaneous aesthetic judgements (of sublimity). There does not seem to be a simple analogue of the fluency-based account in the case of sublimity experiences and judgements.

A more ambitious strategy would try to accommodate the sublime within the scope of the fluency-based account. Bullot and Reber (2013) propose a distinction between perceptual and conceptual fluency and claim that cases of aesthetic experience in which disfluency is involved (which may or may not involve the sublime) are always, eventually, conceptually fluent. For instance, Bridget Riley's paintings are clearly *visually* disfluent and yet *conceptually* fluent to the extent that the mind enjoys reflecting on its own contribution to the perceptual experience of the world. In the specific case of the sublime (which they do not discuss as such), the authors' view could be that our aesthetic experience is at least conceptually fluent, although it is strongly disfluent at lower (perceptual and affective) levels. Indeed, as some of our previous scientific examples illustrate, in having a sublimity experience, the mind enjoys reflecting on the limits of perception, imagination and standard frames of reasoning. As Keltner and Haidt (2003) put it in Piagetian terms, sublimity experiences involve a "need for accommodation", i.e., an urge to go beyond standard frames of reference insofar as they have become useless at dealing with the greatness of the sublime.

The ambitious strategy might underestimate the role of disfluency as a causal determinant of spontaneous aesthetic judgements. Disfluency is not a dispensable component of the whole affective experience that is supposed to feed such judgements, over which fluency eventually wins. As was noted above, fluency feeds judgements of beauty only if the subject is *ignorant* of the source of fluency. More precisely, the subject may form a judgement of beauty if she feels fluency but also some uncertainty as to the source of the fluency. The latter feeling of uncertainty involves processing *disfluency*. Thus, the whole affective experience that feeds the subject's spontaneous judgements of beauty involves both fluency and disfluency (see Dokic 2016). In the case of the sublime, the disfluency is presumably stronger, since it is associated with a disturbing limit-experience, in which the mind faces some of its own cognitive boundaries and limitations. Yet, the mind somehow overcomes or accommodates this limit-experience, giving rise to a pleasurable experience. In a nutshell, any aesthetic experience, whether about something beautiful or sublime, involves a proper balance of fluency and disfluency, compatible with overall aesthetic pleasure.

With this admittedly schematic picture of aesthetic experience in mind, let us come back to the claim that aesthetic considerations guide the scientific practice. We have seen how a subset of these considerations, namely those which pertain to beauty and its symptoms (such as simplicity, elegance, symmetry, etc.) may guide the scientist to judgements of truth and understanding. What about aesthetic considerations having to do with the sublime? Can they guide the scientific practice too? If so, what are the non-aesthetic judgements that they motivate?

We have already stressed (section 3) the importance given by scientists to values such as complexity and irregularity. It is interesting to note that when he comments on the role of beauty in science, Poincaré seems to point to a distinction between two types of aesthetic considerations:

> And it is because simplicity, because vastness, is beautiful that we seek by preference simple facts and vast facts, that we take delight, now in following the giant courses of the stars, now in scrutinising with a microscope that prodigious smallness which is also a vastness, and now in seeking in geological ages the traces of a past that attracts us because it is far away.
>
> (Poincaré 1908: 16)

In this passage, Poincaré observes how the scientist is motivated to study "vast facts" [les faits grandioses], even though cognition of greatness typically involves disfluency rather than fluency. Great facts are mathematical, spatial, temporal or dynamic facts that defy our standard schemes of thought and reasoning.[9] Again, great facts call for accommodation, and challenge the scientist to find radically novel ways of theorising about the world.[10]

One can surmise, then, that spontaneous judgements of sublimity can enter heuristics, which draw the scientist's attention to highly challenging phenomena and domains of enquiry. Such judgements can also contribute to our spontaneous evaluation of a theory as innovative. It does not follow that processing disfluency by itself is relevant to scientific practice. As we have observed, too much disfluency signals confusion, and diverts rather than draws and maintains attention. However, processing disfluency against a background of processing fluency may be the symptom that we have arrived at a theory whose explanatory value is radically novel while the components of the theory are familiar.

We have suggested that the objects of sublimity experiences are more relational than the objects of beauty experiences. While the latter seem to point to intrinsic features of the beautiful objects, the former are essentially about our own cognitive relationship with the world. The relational structure of sublimity experiences can then provide another route to understanding.[11] We have seen how values such as simplicity, unity, elegance, etc., which are correlated to fluency at the psychological level, can guide the scientist to judgements of understanding (which may or may not be accompanied by judgements of truth). We now see that aesthetic considerations pertaining to sublimity can also motive the scientist to form other judgements of understanding, which involve the feeling that the theory or set of hypotheses under consideration has been pushed towards the limits of what we may cognitively encompass as human beings. This is what happens when, for instance, we grasp the deep meaning of General Relativity. Sublimity experiences are limit-experiences, but limit-experiences also play a role in the pursuit of understanding, which again may point to a common psychological core.

5 Conclusion

In this chapter, we have tried to clarify the relevance of aesthetic considerations to science. We have distinguished three levels at which such considerations might apply: the level of the objects of scientific inquiry, the level of the scientific constructions (theories, hypotheses, models, etc., but also experimental settings), and the level of the scientific practice (constructing and evaluating theories, etc., and designing and conducting experiments).

We have endorsed aesthetic pluralism, in particular the idea that the aesthetic domain encompasses not only the beautiful, but also the sublime. The aesthetic category of the sublime has been largely neglected in discussions of the relationship between aesthetics and science, but we have shown that scientists themselves have often pointed to aesthetic properties and values pertaining to the sublime. Vast facts and thought-provoking theories might trigger sublimity more than beauty experiences. We have given several illustrations of how aesthetic experiences and judgements, about scientific objects or constructions, guide the practice of scientists in different ways, depending on whether they belong to the beautiful or the sublime.

The claim that aesthetic considerations can guide the scientific practice might be grounded on deep ontological connections between aesthetic and epistemic values, having to do with truth, justification and understanding. Independently of such connections, about which we can remain relatively neutral here, this claim can also be grounded on a theory of aesthetic experience as being constituted by epistemic feelings and emotions. Aesthetic experience has to do with surprise, interest, curiosity, perplexity, etc., which are variations on the themes of familiarity and novelty. An aesthetic object cannot be too familiar, on pains of being boring, and it cannot be too novel, on pains of being confused. Arguably, the very same epistemic values pervade the scientific practice, and guide the scientist to the elaboration and evaluation of a theory whose conclusion may be novel while the building-blocks of the theory are familiar. As the beautiful and the sublime involve different blends of familiarity and novelty, it is not an accident that the complex feelings underlying spontaneous judgements of beauty and sublimity play differentiated roles in the epistemic heuristics used by the scientist. Some aspects of a theory may strike us as being beautiful because of the way it articulates something with which we were already familiar, and other aspects of the theory may strike us as being sublime because it points to something entirely novel and deep. Familiarity and novelty form the common core of aesthetic and epistemic values, and this explains why considerations that are aesthetically relevant can also be epistemically relevant to theory building and evaluation.

More specifically, we have shown that aesthetic considerations can feed not only judgements about the truth or correctness of scientific theories (hypotheses, etc.), but also judgements of understanding. Although beauty can guide the scientist to some judgements of understanding having to do with the internal structure of theories, we have suggested that sublimity may be responsible for further judgements about the limits of human understanding. More generally, we surmise that the relational nature of sublimity experiences makes them apt as a guide to relational knowledge, which seems crucial for scientific understanding.

Acknowledgements

We thank Steven French and Milena Ivanova for their critical and constructive comments. We are also grateful to the audience at the "Aesthetics of Science" Conference (held in Leeds) for its valuable observations. This research has been supported by the SublimAE Project (ANR-18-CE27-0023-01), and by the ANR-17-EURE-0017 FrontCog and the ANR-10-IDEX-0001-02 PSL.

Notes

1. On a strong version of aesthetic pluralism, there are at least two distinct *kinds* of aesthetic experiences and properties, such as beauty and sublimity. Weaker versions allow for a continuum between beauty and sublimity, with typical instances of each. Here we stress different cognitive patterns of aesthetic experiences and corresponding properties while being relatively neutral on which version of aesthetic pluralism is correct.
2. Ivanova (2017b) stresses three possible different roles played by aesthetic considerations in scientific enquiry, namely a motivational role, a heuristic role and an epistemic role. In section 4 we will be mainly concerned with the latter two roles.
3. It is often noted that sublimity experiences involve reverence or admiration, potentially due to their overwhelming aspect (see Keltner and Haidt 2003). That is true, but it might be argued that beauty experiences also involve feelings of elevation or reverence, though they do not show an overwhelming

aspect. Prinz (2011) stresses the importance of these kinds of feeling in aesthetic experience in general.

4. Judgements of beauty have been repeatedly reported to activate the medial prefrontal cortex, encompassing both the medial orbitofrontal cortex (mOFC) and the rostral anterior cingulate cortex (rACC), for paintings (e.g., Kawabata and Zeki 2004; Ishizu and Zeki 2011), music (Ishizu and Zeki 2011). The pivotal role of the mOFC has been supported by a meta-analysis of neuroimaging studies (Brown, Gao, Tisdelle, Eickhoff and Liotti 2011). Indeed, positive aesthetic judgement about stimuli from different sensory modalities (vision, audition, gustation and olfaction) recruited partially overlapping portions of this brain region.
5. The following passage from John F. W. Herschel illustrates the role of wonder in the scientist's experience but also its self-sustaining character: "Accustomed to trace the operation of general causes and the exemplification of general laws, in circumstances where the uninformed and unenquiring eye *perceives neither novelty nor beauty, he walks in the midst of wonders* (. . .). Nor is it a mere passive pleasure which is thus communicated. A thousand questions are continually arising in his mind, a thousand subjects of enquiry presenting themselves, *which keep his faculties in constant exercise, and his thoughts perpetually on the wing*" (Herschel 1851: 15—italics ours).
6. Although their study undoubtedly offers important insights into the nature of the sublime, Ishizu and Zeki's findings should be taken with caution for two main methodological reasons (see Arcangeli, Dokic and Sperduti 2018). First, the study's experimental setting is fragile: pictures of natural scenes presented on a computer screen within a scanner are clearly limited in their grandeur and capacity as triggers of sublimity experiences. Second, the authors' comparison with the case of beauty appeals to a quite different experimental material, involving paintings rather than natural scenes.
7. The *Garganta del Diablo* is a set of waterfalls (80m high) that fall into a narrow canyon, which concentrates the largest flow of the Iguazu Falls, being in turn (located on the Argentina-Brazil border) the largest flow in the world.
8. Kant's theory has been considered, for instance, as "radically subjective" (Shapshay 2014: 96) or as an "egoistic model" of the sublime, which ignores "the distinctly other-directedness of the sublime experience" (Cochrane 2012: 13). Tom Cochrane puts forward a model in which sublimity experiences are primarily object-centred and involve a sense of self-negation.
9. The original French terms "grandeur" and "grandiose" are differently translated in English. "Vastness" and "vast" are employed by Francis Maitland in his translation of *Science and Method* (1914, London: T. Nelson). "Grandeur" and "sublime" are preferred by George Bruce Halsted (see *The Foundations of Science: Science and Hypothesis, The Value of Science, Science and Method*, 1913, New York: The Science Press).
10. See again Herschel's quotation in footnote 5.
11. See also Ivanova (2017a). We take on board her claim that Poincaré links beauty to scientific understanding (i.e., revealing "hidden kinships" and "real relations" in the phenomena). We are suggesting in addition that the sublime may be relevant too, and that Poincaré seemed to acknowledge the distinction between beauty and sublimity.

Bibliography

Addison, J. (1712), 'Paper II. On the Pleasure of the Imagination', in J. Addison and R. Steele (eds.), *The Spectator*, 412, Monday, June 23.

Allport, F. H. and Lepkin, M. (1945), 'Wartime Rumors of Waste and Special Privilege: Why Some People Believe Them', *Journal of Abnormal and Social Psychology*, 40: 3–36.

Arcangeli, M., Dokic, J. and Sperduti, M. (2018), 'The Beautiful, the Sublime and the Self', in F. Cova and S. Réhault (eds.), *Advances in Experimental Philosophy of Aesthetics*. Bloomsbury, pp. 175–196.

Brady, E. (2013), *The Sublime in Modern Philosophy: Aesthetics, Ethics, and Nature*. Cambridge University Press.

Breitenbach, A. (2013), 'Aesthetics in Science: A Kantian Proposal', *Proceedings of the Aristotelian Society*, 113: 83–100.

Brown, S., Gao, X., Tisdelle, L., Eickhoff, S. B. and Liotti, M. (2011), 'Naturalizing Aesthetics: Brain Areas for Aesthetic Appraisal across Sensory Modalities', *Neuroimage*, 58: 250–258.

Bullot, N. J. and Reber, R. (2013), 'The Artful Mind Meets Art History: Toward a Psycho-Historical Framework for the Science of Art Appreciation', *Behavioral and Brain Sciences*, 36: 123–137.

Burke, E. (1759), *A Philosophical Enquiry into the Origin of Our Ideas of the Sublime and the Beautiful*, 2nd ed., ed. J. T. Boulton. University of Notre Dame Press, 1968.

Cochrane, T. (2012), 'The Emotional Experience of the Sublime', *Canadian Journal of Philosophy*, 42: 125–148.

Cova, F. and Réhault, S. (2018), *Advances in Experimental Philosophy of Aesthetics*, Bloomsbury.

Dokic, J. (2016), 'Aesthetic Experience as a Metacognitive Feeling? A Dual-Aspect View', *Proceedings of the Aristotelian Society*, 116: 69–88.

Einstein, A. (1932), *Mein Glaubensbekenntnis*. September/October recording.

Einstein, A. (1934), *Mein Weltbild*. Rudolf Kayser. [1935 en. trans. by Alan Harris, London: John Lane The Bodley Head].

Faraday, M. (1932–36), *Faraday's Diary*, ed. T. Martin, 7 vols. London: Bell.

Feynman, R. (2005), *Perfectly Reasonable Deviations From the Beaten Track: The Letters of Richard P. Feynman*, ed. M. Feynman. Basic Books.

Feynman, R., Leighton, R. B. and Sands, M. (1964), *The Feynman Lectures on Physics*, Vol. I, Mainly Mechanics, Radiation and Heat. Online edition at www.feynmanlectures.caltech.edu/

Gould, S. J. (1991), *Bully for Brontosaurus: Reflections in Natural History*. Norton.

Herschel, J. F. W. (1851), *Preliminary Discourse on the Study of Natural Philosophy*. Longman, Brown, Green & Longmans.

Ishizu, T. and Zeki, S. (2011), 'Toward a Brain-Based Theory of Beauty', *PLoS One*, 6: e21852.

Ishizu, T. and Zeki, S. (2014), 'A Neurobiological Enquiry into the Origins of Our Experience of the Sublime and Beautiful', *Frontiers in Human Neuroscience*, 8: 891.

Ivanova, M. (2017a), 'Poincaré's Aesthetics of Science', *Synthese*, 194: 2581–2594.

Ivanova, M. (2017b), 'Aesthetic Values in Science', *Philosophy Compass*, 12: e12433. https://doi.org/10.1111/phc3.12433

Kant, I. (1790/2000), *Critique of the Power of Judgement*, ed. P. Guyer, trans. P. Guyer and E. Matthews. Cambridge University Press.

Kawabata, H. and Zeki, S. (2004), 'Neural Correlates of Beauty', *Journal of Neurophysiology*, 91: 1699–1705.

Keltner, D. and Haidt, J. (2003), 'Approaching Awe, a Moral, Spiritual, and Aesthetic Emotion', *Cognition and Emotion*, 17: 297–314.

Levinson, J. (2011), 'Beauty Is Not One: The Irreducible Variety of Visual Beauty', in E. Schellekens and P. Goldie (eds.), *The Aesthetic Mind: Philosophy and Psychology*. Oxford University Press, pp. 190–207.

Poincaré, H. (1908), *Science et Méthode*. Paris: E. Flammarion.

Prinz, J. (2011), 'Emotional and Aesthetic Value', in E. Schellekens and P. Goldie (eds.), *The Aesthetic Mind: Philosophy and Psychology*. Oxford University Press, pp. 71–88.

Reber, R. (2012), 'Processing Fluency, Aesthetic Pleasure, and Culturally Shared Taste', in A. P. Shimamura and S. E. Palmer (eds.), *Aesthetic Science: Connecting Minds, Brains, and Experience*. Oxford University Press, pp. 223–249.

Russell, B. (1919), *Mysticism and Logic and Other Essays*. Allen and Unwin.

Schaeffer, J.-M. (2015), *L'Expérience Esthétique*. Gallimard.

Schopenhauer, A. (1819/1844), *Die Welt als Wille und Ausstellung*, translated as *The World as Will and Representation*, Vol. I. by E. F. J. Payne (1969). New York: Dover Publications.

Schwarz, N. (2018), 'Of Fluency, Beauty, and Truth: Inferences From Metacognitive Experiences', in J. Proust and M. Fortier (eds.), *Metacognitive Diversity: An Interdisciplinary Approach*. Oxford University Press, pp. 25–46.

Scruton, R. (2009), *Beauty*. Oxford University Press.

Shapshay, S. (2014), 'The Problem and Promise of the Sublime: Lessons From Kant and Schopenhauer', in J. Levinson (ed.), *Suffering Art Gladly: The Paradox of Negative Emotion in Art*. Palgrave Macmillan, pp. 84–107.

Todd, C. S. (2008), 'Unmasking the Truth Beneath the Beauty: Why the Supposed Aesthetic Judgments Made in Science May Not Be Aesthetic at All', *International Studies in the Philosophy of Science*, 22: 61–79.

Zajonc, R. B. (1968), 'Attitudinal Effects of Mere Exposure', *Journal of Personality and Social Psychology: Monograph Supplement*, 9: 1–27.

Zeki, S., Romaya, J. P., Benincasa, D. M. and Atiyah, M. F. (2014), 'The Experience of Mathematical Beauty and Its Neural Correlates', *Frontiers in Human Neuroscience*, 8: 68.

7 How Can Loveliness Be a Guide to Truth? Inference to the Best Explanation and Exemplars

Alexander Bird

1 Introduction—Inference to the Best Explanation and Its Problems

Science makes frequent use of Inference to the Best Explanation (IBE). While IBE is not the only form of inference to be found in science, it is key to our assessments of the plausibility of many theoretical scientific claims. The best discussion of what IBE is and whether it can give us rational beliefs and epistemic preferences concerning theories remains Peter Lipton's (2004) *Inference to the Best Explanation*.

IBE is about choosing among explanations. It is a matter of choosing among *potential* explanations of some phenomenon the one that is the best by certain criteria. If there is a suitable best potential explanation, IBE says that we may infer that it is the *actual* explanation, i.e. that the explanatory hypothesis is true.

According to Peter Lipton, IBE is a two-stage process, where both stages are filters of potential explanations (Lipton 2004: 56–64):

Stage 1: The first stage filters out the implausible explanations. The imaginative capacity of scientists generates all the plausible potential explanations and just leaves the remainder unconsidered.

Stage 2: At the second stage, scientists investigate the live potential explanations that have passed through the first filter, and ultimately rank them according to their explanatory goodness, in order to select the top ranking explanation as *the* explanation.

What I have called 'explanatory goodness', Lipton calls 'loveliness'. I shall take 'loveliness' to refer to the good-making features of an explanation, whatever they may be. On that construal it is almost trivial that when people make an inference to the best explanation, they are inferring the loveliest explanation. But it is far from trivial to know what loveliness is. The following are substantial questions: What are the lovely making features of an explanatory theory? And how do scientists come to see and respond to these features in a theory? Lipton himself thinks that loveliness is a matter of the understanding that a potential explanation would give, if it were true. Others have different views. According to the view I articulate below, loveliness is context-dependent (which is one reason why we find it difficult to agree on a single, fixed account of what it is).

Both stages in IBE raise important philosophical questions. A crucial problem concerns the first stage.

> **Underconsideration** The actual explanation may not be among the potential explanations we consider and rank. If so, the top ranked explanation will not be true. The truth was not even considered.

IBE cannot generate knowledge if we do not consider and assess all prima facie plausible explanations. Since stage 1 filters out so many logically possible explanations, what confidence can we have that the actual explanation is allowed through? The stage 2 ranking is of no use if the actual explanation hasn't made it through stage 1 on account of the scientists' failure to think of it. Why should we think that we have considered all the relevant possibilities? Kyle Stanford (2006) argues that we do in fact have reason, as demonstrated by the history of science, for thinking that our reasoning about novel theoretical claims suffers from what he calls 'the problem of unconceived alternatives'. We can articulate the problem as a problem about the truth-directedness of scientists' imaginations. The very idea of IBE concedes that the scientific imagination is not directed unfailingly on the truth. Multiple hypotheses—products of the imagination—are considered, of which at most one is true. So, mostly, the scientific imagination produces false hypotheses. Given that the scientific imagination does not have a strong link with the truth, we ought not expect it to have the magical ability always or even very frequently to generate the true hypothesis alongside the false ones it generates.

Assuming that the actual explanation *is* among those investigated at stage 2, two problems immediately raise their heads, which Lipton calls 'Hungerford's objection' and 'Voltaire's objection'.

> **Hungerford's objection** Loveliness, being an aesthetic quality or akin to one, is subjective. If loveliness is subjective, then it cannot correlate with the truth, which is objective. Hence the loveliness of a potential explanation cannot be an indicator its truth.

Lipton here borrows Margaret Hungerford's line in *Molly Bawn* that 'beauty is in the eye of the beholder'. As Kragh (1990: 287) puts it, 'The main problem is that beauty is essentially subjective and hence cannot serve as a commonly defined tool for guiding or evaluating science'. It is unclear whether explanatory loveliness is itself an aesthetic quality. But even if it isn't, it is prima facie very plausible that it is akin to an aesthetic property at least as regards subjectivity, for four reasons. First, we tend to use *aesthetic terminology* to describe the good-making features of explanations and theories, whether Lipton's 'loveliness', or McAllister's (1999) 'beauty', or Glynn's (2010) 'elegance'. The second reason for thinking that loveliness is an aesthetic quality or is at least a subjective quality is that it is *ineffable*. It is difficult to say exactly what makes an explanation lovely. We are better able to recognise it than we are able to articulate what it is. Aesthetic qualities are ineffable in this way—many people will know that Mozart's oboe quartet is beautiful but cannot say in virtue of what properties it is beautiful. The ineffability of explanatory loveliness may in part explain our use of aesthetic terminology. The third reason is also behind the use of aesthetic terminology to describe theories. We take a certain kind of *pleasure* on grasping a particularly revealing, unifying, apt, or economical explanation. This pleasure is the same as or is very similar to the kind of pleasure one has on experiencing (with understanding) an artistically valuable poem, sonata, or still life. Finally, we find *variability* across different disciplines or across different times in the same discipline, regarding what counts as lovely. Even when we can agree that something such as simplicity is an aspect of loveliness, simplicity looks to be a different thing in physics from in physiology. Even within one field, astronomy, what counted as simplicity in one era (circular motion until the time of Kepler) was later no longer a determinant of loveliness. This variation in assessments of loveliness is what one would expect if loveliness is or is like an aesthetic quality, subject to changes in taste.

Voltaire's objection If the loveliest explanation is to be true, then the actual world has to be lovelier than other possible worlds. But we should not expect the actual world to be the loveliest.

This objection says that the IBE enthusiast has an unjustified Panglossian faith that the actual world is the loveliest of all possible worlds. Even if loveliness is objective, there will be many worlds where it does not correlate with truth. So why think that loveliness and truth correlate in ours? Lipton regards Voltaire's objection as a version of Hume's problem of induction: there are worlds where apparent regularities are not true regularities, so why think that the apparent regularities are true regularities on our world? Voltaire's problem may be even worse. For induction to be a useful guide to the truth, it need only be that the actual world has a high enough degree of regularity. It does not have to be the most regular of worlds. Voltaire's problem is more demanding. For IBE to lead frequently to the truth, it has to be that the actual world is the loveliest of possible worlds, or at least among the most lovely worlds. And that just seems implausible.

In other work (Bird 2020) I argue that Lipton's own solutions to these problems are only partially satisfactory and that the sceptical worries the three objections raise remain. In this chapter I provide my own response to those objections. That response appeals to the idea that a community's standards of explanatory goodness are acquired in the process of scientific training and learning which uses *exemplars*, drawing on a proposal by David Walker (2009, 2012).

2 Exemplars

In this section I articulate what is meant by an 'exemplar' and explain why thinking with exemplars is an important part of cognition in science. While the exemplar idea originates with Kuhn, its importance extends well beyond its place in Kuhn's (earlier) philosophy of science.

2.1 Exemplars in Kuhn's Thought

In the 'Postscript 1969' to the second edition of *The Structure of Scientific Revolutions* Thomas Kuhn explicates in detail what he means by the contentious notion of 'paradigm'. There is a sense of 'paradigm' concerning which Kuhn (1970: 10) remarked, 'By choosing it, I mean to suggest that some accepted examples of actual scientific practice—examples which include law, theory, application, and instrumentation together—provide models from which spring particular coherent traditions of scientific research'. To express this meaning of 'paradigm' Kuhn also used the term 'exemplar', thereby distinguishing paradigms-as-exemplars from paradigms in a broader, more sociological sense. Exemplars are, Kuhn (1970: 187) said, the most novel and least understood aspect of his book. Exemplars are social in that they are shared exemplary illustrations and puzzle-solutions, and it is by being shared that they can explain the development of a social activity such as science. But the key innovative feature of the exemplar idea is not so much the social as the psychological function of exemplary puzzle-solutions.

Kuhn's central claims about the functioning of exemplars are as follows:

- (i) the processes of scientific cognition are driven by perceived similarity to exemplary puzzles and their solutions (exemplars);
- (ii) the ability to perceive such similarities is acquired by training with exemplars;
- (iii) that ability is primarily an ability to recognise patterns and relevant similarities;
- (iv) it is therefore not an ability that is mediated by following rules.

Focusing on (i), the principal processes of scientific cognition in question are:

(a) selecting puzzles;
(b) solving puzzles;
(c) assessing the quality of a proposed puzzle-solution.

It had been widely thought that in order for science to be rational scientists must follow certain rules of rationality, at least in the so-called context of justification. Optimally, philosophers of science should be able to reconstruct the a priori rules of scientific method by which our theories are justified and hence accepted. Kuhn himself rejected the sharp distinction between the context of discovery and the context of justification since exemplars play a role in both (see (i) above). Moreover, justification and the explanation of scientific change do not depend on scientists following, even unconsciously, the rules of scientific method. Rather, what drives theory-choice is perceived similarity to an exemplar. This is most clear during normal science but is true also even in revolutionary science.

Kuhn (1970: 36) compared normal science to solving puzzles such as crossword puzzles. One feature of puzzle-solving to which Kuhn sought to draw our attention is the fact that one can learn to do crossword puzzles more quickly and easily simply by practising them. Although one can write very effective computer programmes that play chess by implementing an algorithm, good human players recognise patterns in positions in a non-algorithmic way that is acquired by past exposure to many similar positions. Kuhn (1970: 47) also likens practice with scientific puzzles to finger exercises on a musical instrument, emphasising that the use of exemplars to solve problems is rather more like a skill than like the explicit application of rules.

The most obvious cases of learning by repeated exposure and practice in science will be found in the mathematical sciences where students learn from a textbook and lectures and are given problems to solve. The easier problems will be very similar to the problems that are solved in the textbook and lectures. Harder problems will be less obviously similar. Practice with the former will give a student an improved ability that will in due course allow them to tackle the more difficult problems. As suggested, what makes a difficult problem more difficult will be not just increased complexity but also the fact that it is less clearly similar to a problem the student has seen before. What, according to Kuhn, repeated practice with exemplary puzzle solutions provides is an ability to spot a similarity between a new puzzle and one that the student has seen before.

We are familiar with the idea of a non-rational power of recognising similarity in the perceptual sphere, for example, recognising the similarity between faces of members of the same family, or hearing that two tunes are similar. It is also true that we can recognise similarities between more abstract patterns and structures and this similarity recognition is likewise a non-rational one.[1] Of course, to say that this capacity is non-rational is not to say that it is irrational. It is clearly not irrational to be able to spot the similarity between a mother and her daughter. Rather what is meant is that this capacity is not mediated by a process of following rules.

Kuhn's purposes in drawing our attention to the existence and function of exemplars are (i) to undermine the then prevailing logical empiricist view that theory choice is governed by a priori rules of rationality (a view exemplified by Carnap's inductive logic or Popper's falsificationism); (ii) to explain certain 'revolutionary' episodes of theory-change—these involve changes in the governing exemplars; and (iii) to explain an interesting aspect of such changes, namely incommensurability. In fact Kuhn himself, from the mid-1970s, seemed to lose interest in the exemplar idea and instead focused on more philosophical accounts of incommensurability. At the same time, anti-realist ideas, absent from the first (1962) edition of *The Structure of Scientific Revolutions*, came more to the fore.

2.2 Exemplars Beyond Kuhn

The value of the exemplar idea can be divorced from Kuhn's later (albeit moderate) anti-realism. Indeed, it is entirely consonant with an approach to scientific cognition that is naturalistic and realist in that it holds that we are capable of latching onto the truth and that we have this capacity in virtue not of grasping a priori rules of rationality but in virtue of possessing certain reliable modes of thinking that themselves may be innate or may be the products of experience and learning. Such capacities are therefore themselves open to scientific investigation and evaluation.[2] Facial recognition, as just mentioned, is a simple example of such a capacity, largely an innate capacity. The ability to assess the strength and potential of a chess position or the ability to recognise the composer of a previously unheard piece of music are learned capacities of this sort. Kuhn regards the ability to see the similarity between a new scientific problem and a previously solved one as another learned recognitional capacity. Kuhn's idea has been taken up in this spirit by a number of philosophers (e.g. Andersen, Barker and Chen 1996; Nickles 2003), in particular those who link the idea of exemplar to a wider capacity for pattern recognition and analogical thinking (e.g. Margolis 1987; Bird 2005). At the same time, there is evidence from more recent cognitive psychologists (Leake 1998; Dunbar 1996, 1999; Gentner, Holyoak and Kokinov 2001) that confirms the view that scientists as well as others do reason by drawing analogies between the problems they face and previously solved problems (i.e. exemplars).

According to this research, scientists look for solutions to their scientific problems by looking for analogies between their problems and ones that have been solved before. Indeed a scientist will see the world through the lenses of her exemplars, in the sense that she will classify and conceptualise the world without the mediation of an inference. For example, the treehopper *Cyphonia clavata* has an extraordinary growth that looks very much like a tree ant. An evolutionary biologist will *immediately* see this as an adaptation, without going through a thought process involving 'these are the reasons . . ., *a*, *b*, *c*, . . . why this body-structure is most plausibly an adaptation to deter predation'. A pre-Darwinian naturalist would not have conceptualised what he saw as an adaptation, but might have, again without the mediation of inference, seen it as an instance of Divine Providence. A consequence of the fact that cognition using exemplars can be a wholly or partly unconscious process, like facial recognition, is that the subject can fail to be aware of all the factors influencing the cognitive process. It may simply not be possible to give a complete reconstruction of the process in terms of a rational inference. So such cognitive capacities are like intuition in that they are partly or wholly unmediated by conscious reasoning. On the other hand, they are unlike intuition in that they are learned. They are 'second-nature', or, more prosaically, 'quasi-intuitive cognitive capacities' as I (2007) call them. And they are plausibly what Duhem (1914) had in mind when he talked of the 'good sense' that scientists use to prefer one theory to another.[3]

Cognizing with exemplars is consistent with realism. An apt analogy with an exemplar that is the correct solution to a problem can be conducive to finding the correct solution to a new problem. But this consistency with realism does not exclude the possibility of systematic error. If the analogy is inapt or the exemplar is itself an error, then thinking with exemplars will lead to false beliefs (I give some examples below). So this theory of scientific cognition may permit realism but it certainly does not guarantee it. However, in the context of defending IBE from the three sceptical problems—Underconsideration, Hungerford's problem and Voltaire's problem—that, I will argue, is enough.

2.3 Exemplars and Loveliness

We are now in a position to see how the concept of an exemplar can provide a solution to the three problems facing IBE. The key idea in what follows is that scientists acquire their standards of explanatory goodness thanks to their exposure to exemplars. Although I introduced the idea by referring to Kuhn, the basic idea goes back to Aristotle, that we acquire our values through certain kinds of training, rather than by the learning of rules. We acquire virtuous dispositions and learn how to recognise the virtues in others by moral training, by witnessing the virtuous actions of others and imitating such actions, repeatedly acting as the virtuous person does. In science, training with exemplars is similar: the young scientist engages in academic exercises of problem solving with examples—from textbook questions to apprenticeship as a doctoral student. The young scientist thereby acquires the capacity to see new problem situations as similar to exemplary ones, to deploy theoretical and experimental tools in an analogous fashion, and to judge others' attempts to do the same. In particular she acquires the capacity to see theoretical problem situations as calling for a certain kind of explanation and correspondingly learns to judge how well a proposed explanation meets the demands of some problem situation. In a related vein Walker (2012) provides a paradigm-focused understanding of Lipton's notion of explanatory loveliness.[4]

The discovery of Coulomb's law provides an illustration of the effect of exemplars on scientific cognition and on judgements of explanatory goodness in particular. In the context of Newtonian mechanics and the success of Newton's law of gravitation it was natural for a physicist confronted with an unstudied source of force to want to find its law. So the exemplar of gravitation generates the problem for electrostatics. Furthermore it directs us towards a solution: by analogy with Newton's law, an inverse square law naturally suggests itself; indeed Coulomb was not the first to propose an inverse square law. Franz Aepinus had proposed it without any experimental evidence and Charles Stanhope attempted a demonstration that was widely accepted despite being flawed. Coulomb employed a torsion balance in his experiment, much the same, though differing in scale, as the torsion balance devised by John Michell and which Henry Cavendish used 'to weigh the Earth'. Coulomb's experiment was not universally considered to have proven his case: other scientists had difficulty in replicating his results and some proposed different laws on the basis of results using the same apparatus; modern scholars differ on whether Coulomb really could have achieved the results he claimed he did (Heering 1992, 1994; Falconer 2004; Martínez 2006). Nonetheless, Coulomb's proposed law was accepted by the majority, despite its empirical shortcomings. It is apparent that as far as Coulomb and many of his contemporaries were concerned, the very fact of the analogy with the law of gravitation played a powerful role in influencing their judgement in accepting the law. That role cannot be thought of as an evidential role: it is difficult to see how the fact that Newton's law is an inverse square law is *evidence for* Coulomb's law, particularly in the context of IBE, since the form of Newton's law is not explained by Coulomb's law having that form (nor vice versa). Nor would it be true to the psychology of the scientists; the evidence they appealed to was what was known of electrical phenomena, in particular the various experimental results. Rather, the common form of Coulomb's proposed law and Newton's well-attested older law made the former a better explanation of the evidence.

In the following sections I argue that this exemplar-based account of our judgements of explanatory goodness resolves the three objections to IBE. (As I have emphasised, to respond satisfactorily to these objections is to remove prima facie reasons to reject IBE and so scientific realism; they are not intended to be arguments for the conclusion that one *must* rationally accept IBE and realism.) The key features of judgements of explanatory goodness, being based on learning with exemplars, are:

The judgements are not entirely a matter of rational, conscious reflection. Since they are instances of recognising patterns and similarities, there will be an unconscious element to the cognitive process.

Consequently, the subject will be able to make a judgement without being able to articulate fully the basis on which the judgement is made—they may be able to say little more than 'it feels right' or 'that is how it strikes me'.

The factors that do influence judgements will be dependent on the nature of the exemplars used in learning. Hence standards of explanatory goodness will not be a priori and may be specific to the discipline or sub-discipline from which the exemplars come, and to their stage of development.

3 Exemplars and Hungerford's Objection

Hungerford's objection says that our evaluative standards are too subjective to correlate with truth. Loveliness looks very much like an aesthetic quality, and therefore subjective.

We should note that 'subjective' can be ambiguous. A quality can be subjective in the sense that it is response dependent, where the response is a private state. And it can be subjective in the sense of its holding is merely a matter of opinion. These will often go together. 'Chacun à son goût'—we might think that what food we find delectable and which unpleasant is usually just a matter of taste, and there is no objective deliciousness or disagreeableness that these could correspond to. On the other hand, some qualities are subjective in the first sense but not the second—colour experiences are, arguably, like that. It is common to think that aesthetic qualities are of the type that are subjective in both senses: all there is are our inner aesthetic experiences; there is no fact about their being correct, aesthetic taste being a matter of personal opinion only. The line from Hungerford's *Molly Bawn*, 'beauty is in the eye of the beholder', expresses that view. But Hume and Kant both thought otherwise, holding that there are objective standards of taste to which the subjective aesthetic response of a suitable critic will correspond. Let's call qualities that are subjective in both senses 'merely subjective'.

Various features of explanatory loveliness do suggest that it is a subjective quality and indeed a merely subjective quality like (or among) the aesthetic qualities, at least as Hungerford's line takes them to be. In brief, these features were: i) Aesthetic terminology: writers use subjective qualities to describe explanatory goodness; ii) Ineffability: there is little or no agreement on what explanatory goodness is; iii) Pleasure: we take pleasure in certain explanations, a pleasure that is at least very similar to aesthetic pleasure; iv) Variability: even when we focus on particular qualities that may contribute to explanatory goodness we find that there is variation in what counts as exemplifying these qualities; there is also no rule for weighting the qualities that contribute to explanatory goodness.

The correct response to this objection is to point out that these facts, while prima facie suggestive of mere subjectivity, do not entail it. Indeed they are also what one should expect if the process of learning to produce and evaluate explanations proceeds via training with exemplars rather than through the acquisition of explicit rules.[5] For example, one may learn what prudence is by learning to recognise examples of prudent actions and by taking such actions oneself. Such practical knowledge need not be manifested in an ability to say precisely what features of a given action make it prudent and how they differ from the properties of actions that are judged to be cowardly or avaricious. Prudence may nonetheless be an entirely objective feature of people and their actions. Many cognitive capacities are aimed at determining the presence of some objective property, yet the possessors of such skills are unable to articulate in any detail how they exercise such powers.

For example, a music lover who is not a trained musicologist may be able to identify the composer of a piece of music she has not heard before. She may nonetheless be able to give little concrete justification or explanation of her judgement beyond 'it just sounds like Mendelssohn'. Here we have an example of a cognitive skill detecting an objective property. Other examples include mastery of grammar: most English speakers will regard 'the big red barn' as grammatically preferable to 'the red big barn' without being able to articulate any rule of grammar that they are following in making this judgement—the first just *feels* right, and the second does not (Kihlstrom 1987: 1447). Experts at speed chess will be able to evaluate moves and the resulting chess positions as weak or strong with little or no conscious assessment. Many of these cognitive skills, like the more familiar capacity to recognise a face, are pattern-recognition capacities. In all these cases the detection of the relevant objective property or properties proceeds via a process that is partly or wholly unconscious. Consequently, one would not expect possessors of such capacities to be able to articulate what qualities they are responding to. So what is *not* occurring is the conscious detection of relevant qualities, and then a process of reasoning leading to an assessment. Since the subject cannot articulate some or all of the key determinants of their judgement, that process of judgement will be ineffable to the subject. At the conscious level all that may occur is the response to the problem: a feeling that this is right or good. That feeling will be a positive affect amounting in some cases to pleasure.

In addition, the assessment of goodness may be a matter of balancing a number of criteria. When assessing the comparative strength of two chess positions, a chess master will take into account not only the value of the pieces, but their effectiveness in combination, positional strength, as well as psychological and other factors (e.g. whether the position is of a kind that is well understood or is unfamiliar, whether it favours aggressive or positional play, etc.). So although it may be possible to say what factors go into making a position strong for a player, it will be impossible to say explicitly how those factors should be weighted and combined; that is consistent with strength being objective and the master being able to make an accurate assessment of relative strength of two positions. Kuhn identifies five values (accuracy, consistency, scope, simplicity and fruitfulness) employed in the valuation of a piece of science. But this does not amount to an account of explanatory goodness since there may be no algorithm for weighting these values. While the goodness of an explanation may be an objective property in each case, what makes that explanation good may differ between cases, and may be a complex combination of characteristics. Consequently, the attempt to articulate standards of goodness may be fruitless—the standards are ineffable. Yet training with exemplars may permit a scientist to recognise them nonetheless.

This also explains the 'terminology' feature of judgements of explanatory goodness. If the processes are ineffable and judgements are made according to what feels right, then they will have much the same phenomenal character as aesthetic judgements. In which case it is hardly surprising that terminology is borrowed from aesthetics to describe the subject's response.

A further reason why we might expect it to be difficult to articulate standards of goodness even though those standards are measuring objective features of explanations is that there are different standards for different sciences. Different branches of science are governed by different paradigms: the exemplars of theoretical physics differ from those of inorganic chemistry and those of cell biology. Therefore, on the view being put forward here, there will be different standards of explanatory goodness set in these different disciplines and even within their sub-disciplines. There is not a single set of tacit criteria of explanatory goodness, but multiple sets, and so in that sense there is not a single inferential practice of IBE, but instead there are multiple inferential practices. What counts as 'simple' may not be the same for all scientific disciplines: in theoretical physics it might be the mathematical simplicity of the laws; in physiology it might be the simplicity of a causal process; in evolutionary biology it might be the simplicity of an evolutionary path, and so forth. Explanatory goodness and the general qualities (simplicity etc.) that contribute to goodness will therefore be multiply realisable. Hence one would not expect to be able to articulate a single informative account of what counts as explanatory goodness for all science, even though in each case what is being assessed is objective. Thus variability is what one would expect if standards of goodness are exemplar-driven, given that different branches of science employ different exemplars. Walker (2012: 67) provides a different dimension of variability—over time within a field. Using the example of astronomy he points out that the high explanatory goodness attached to circular orbits was changed to accommodate elliptical orbits. This might appear to a Hungerford objector as a change in taste. We ought rather to see it as a change in explanatory standards that is responsive to the greater success of Keplerian cosmology over Copernican. As a result of that success the former displaces the latter as our preferred exemplar and so as the source of our standards in cosmology.

So the exemplar-based explanation of our ability to evaluate explanatory goodness would predict that even though explanatory goodness is objective, our response to it will have features in common with an aesthetic response and so may appear subjective. That said, I suggest that the exemplar explanation is a better explanation of the data than the hypothesis that our judgements are subjective, because although there is variability between different fields as regards the assessment of goodness, there is rather less variability within fields. The stability of our evaluations of explanatory goodness suggests, though does not prove, that it is indeed an objective feature of explanations. If it were subjective, and genuinely like aesthetic judgements employed in literature or fine art, then one would expect greater variation and disagreement in our judgements. The fact that there is greater agreement in our judgements of explanatory goodness than in our attempts to articulate what it is, suggest that it is an objective property that we come to recognise through learning with exemplars.[6]

4 Exemplars and Voltaire's Objection

In this section I argue that the exemplar approach to understanding explanatory goodness also provides a straightforward response to Voltaire's objection. I then compare this account to James McAllister's 'aesthetic induction'.

4.1 Responding to Voltaire's Objection

Put bluntly, Voltaire's objection suggests that it would be a fluke if, given all the possible worlds there are, the actual world is the best by some given set of standards. It would indeed be a fluke if one were antecedently to lay down a set of standards, and then to ask whether the actual world meets them better than all the others. In effect, that is what happens with our standards of moral and physical good and evil, since those standards are established for reasons that one would not expect to correlate with the way the actual world is. Hence Dr Pangloss's conviction that the world is the best in moral and physical terms seems absurdly optimistic.

But the standards of explanatory goodness are not like that. Our exemplars are chosen for their success. Let us imagine that in a particular mature science, M, success correlates with truth. So in M there will be a number of exemplars that are true or have a high truth-content. These exemplars form the basis by which junior scientists entering M acquire their sense of what a good explanation is. Since standards of explanatory goodness are acquired from successful exemplars, and since in M success correlates with truth, it is perfectly plausible that these standards of explanatory goodness themselves are indicative of truth.

The key to this response is the idea that our standards of explanatory goodness are fixed by features of the actual world—by properties of theories that are actually successful. The exemplar model, put to work in another possible world, would lead to different standards of explanatory goodness, standards that would correlate with the truth in that world.[7] Hence the worry that it would be a fluke if the actual world is the best of all possible worlds explanation-wise can be set aside. The exemplar model tells us that the actual world is the best of all possible worlds (or close to it) *by actual-world standards*—and that is entirely to be expected.

This exemplar-based solution depends on two things:

(i) success does indeed correlate with the truth; and
(ii) in discerning explanatory features of exemplars we are discerning properties that play some role in the success of theories.

Do we know these things to be true?

Before responding to that question, it is important to remark that the exemplar-based approach adopted here does not provide any guarantee that exemplars will set standards that are truth-conducive. A period of normal science may well proceed on the basis of false theories that promote exemplars of explanatory goodness that are not truth-conducive. For example, much medieval science proceeded on the basis that a powerful analogy provides a good explanation. Alchemy employed four elements and four primary qualities. Analogously medicine employed four humours and related these to four temperaments and to personality types as well as the four ages of man. These were all also linked to the four seasons. One can extend the structure so that physiology is explained in terms of four principal organs. One can see how, given such models, parallel explanations in terms of a four-fold division of qualities or kinds, would be seen as explanatorily powerful. Analogies were often religious or quasi-religious. So trial by water was thought to be effective because the accepted best explanation for why the accused floats in water is that water, being an element, is pure and so would reject (and so cause to float) the body of a person who has an impure soul. More recently, the replication crisis in social psychology suggests that its exemplars are not truth-conducive. For example, one highly publicised study (Carney, Cuddy and Yap 2010) maintained that adopting high-power poses causes increased feelings of power and tolerance to risk, alongside changes in biomarkers (raised testosterone and lowered cortisol). This was explicitly modelled on research by Strack,

Martin and Stepper (1988) that showed that contraction of the 'smile muscle' (zygomaticus major) increases pleasure and enjoyment. While both studies found support for their hypotheses, both later failed replication tests. So modelling a new hypothesis on an earlier, successful hypothesis does not guarantee truth and may in fact do the opposite, encouraging falsity.

So, in such cases the tacit criteria of explanatory goodness at work and acquired from exemplars of this sort will not lead to the truth. On the other hand, if the criteria are acquired from exemplars that are true or mostly so, then those criteria *can* be indicative of the truth. Arguably, in some branches of science, such highly truth-like exemplars came to be accepted and so truth-conducive criteria of explanatory goodness came to be employed as a result of the scientific revolution of the seventeenth century. Under such circumstances, further highly-truth-like hypotheses will come to be accepted and will add to the stock of exemplars that in turn reinforce (or gradually refine or develop) the tacit criteria of explanatory goodness.

If one were to show that IBE, as applied in some particular scientific field, really is truth-conducive, then one would indeed have to provide strong arguments for (i) and (ii), that is, we would need to show that we are in the virtuous state of affairs described at the end of the last paragraph. This is the approach taken by Walker (2012: 69). It is to enter into rather more general issues surrounding scientific realism. These are well known and it is not necessary to pursue them here. In responding to Voltaire's problem (and the other two discussed), I am not aiming to convert the sceptic (or even neutral observer) to realism, but to avert a challenge that prima facie would be a reason for the realist to surrender her position. In the light of that goal, it suffices to make the following remarks. So long as (i) and (ii) are in fact true, then IBE will be reliable, and so, according to the epistemological reliabilist, will generate rational belief. In order to rebut scepticism about IBE that is enough (in the reliabilist's eyes). Propositions (i) and (ii) are at least plausible and not prima facie especially improbable. We have defused an objection that, if correct, would have shown that they are false or highly implausible. Lipton is right that we do not need to worry about such general issues of scepticism, and the epistemological responses to them—if on the one hand reliabilism is a satisfactory epistemology, then the current comments suffice; if it is not, then the problems are sufficiently widespread that worrying about the details of IBE is irrelevant. However, in articulating Voltaire's objection, I pointed out that it had an aspect that made it seem especially worrying: the reliabilist solution would be of little avail if the conditions for reliability are ones that seem *very unlikely* to be fulfilled. What the exemplar-based solution does is resolve this additional worry. Contrary to Voltaire's objection, the reliability of IBE does not require the world to be special after all.

4.2 McAllister's Aesthetic Induction

The response to Voltaire's objection that I advocate above has important similarities to what McAllister (1999: 77–81) calls the 'aesthetic induction'.[8] According to McAllister, scientists make aesthetic judgements of theories and together with the application of empirical criteria these play a role in which theories are accepted. This occurs because scientists perceive the presence of certain aesthetic qualities in empirically successful theories and perform an aesthetic induction, projecting the observed correlation of empirical and aesthetic properties. Thus the aesthetic properties can rationally play a role in theory choice, because they are indicators of the empirical properties of theories that scientists value in their pursuit of their scientific goals (e.g. empirical adequacy, truth).

Both this aesthetic induction and the currently proposed idea of exemplar-based judgements of explanatory goodness share the idea that epistemic assessment involves an element that is not purely empirical and which operates by placing an epistemic value on certain qualities of theories that have been found in earlier theories that are regarded as successful. There are significant differences that need to be borne in mind:

The aesthetic induction is straightforwardly an enumerative induction whereas the exemplar-based approach articulates an element of inference to the best explanation.

The aesthetic induction provides a reason to believe a theory that is independent of and parallel to the empirical evidence for the theory, whereas in the exemplar case these are not distinct reasons. An exemplar-based judgement of explanatory goodness contributes to the assessment of empirical evidence. That assessment concerns whether a theory is a good explanation *of the data*, and so tells us whether the data is good evidence for the theory.

The basis of the aesthetic induction is the empirical qualities of prior theories. That the latter are conducive to the goals of science can (allegedly) be known by a priori analysis of those goals. In contrast, the basis of the exemplar-based judgements is just that those theories are held by the community to be exemplars. As mentioned above, those exemplars may be poor guides to achieving the goals of science.

Exemplar-based judgements of explanatory goodness involve largely tacit cognition of the relevant properties whereas McAllister presents the aesthetic induction as if it were an explicit inductive inference. That said, I think McAllister's account would be equally good if the aesthetic induction were tacit.

The aesthetic induction concerns explicitly aesthetic properties whereas exemplar-based judgements of explanatory goodness encompass a wider range of properties. In responding to Hungerford's objection I argued that while aesthetic terminology might often be used, this need not indicate that scientists and others are detecting purely aesthetic properties—if the properties in question were general aesthetic ones such as 'beauty', then we might expect more disagreement than we in fact find. That said, there might be richer, more finely-grained aesthetic qualities on which there is less expert disagreement.[9] And it should be noted that McAllister's understanding of an 'aesthetic' quality is liberal, and includes fit with metaphysical commitments, which will also be part of exemplar-based judgements.

These differences tell in favour of the exemplar-based approach. The aesthetic induction assumes that we can clearly distinguish empirical from non-empirical (e.g. aesthetic) criteria and that we can assess a theory's merits with regard to each separately. While in some cases there may be aesthetic qualities of a theory that can be assessed independently of evidence, more generally it is not possible to make such a distinction, at least if explanationist considerations are central to our practices of theory confirmation.[10] For according to explanationist views (such as IBE) how well evidence *e* supports theory *T* depends on how good an explanation of *e* is provided by *T*. So we cannot assess a theory on empirical grounds (how well the evidence supports it) independently of assessing it on supposedly non-empirical grounds (its explanatory goodness).

5 Exemplars and the Problem of Underconsideration

One of the functions of an exemplar is to help us solve puzzles by generating solutions. So when a puzzle requires an answer in the form of an explanation of certain evidence, exemplars will help us generate potential explanations. The problem of Underconsideration questions whether we are likely to generate the true explanation among the potential explanations we consider. Exemplars solve this problem as follows. We are faced by a new puzzle, P, with data that requires explanation. Let us assume (a) that, frequently, relevantly similar puzzles have similar (true) solutions; and (b) the extant, accepted exemplars in this field are themselves frequently true explanations of their data. Our exemplar-based approach tells us that we seek solutions to P by looking first for past scientific problems to which we have accepted solutions. We then model our potential solutions to P on these accepted solutions—this process might generate more than one potential solution. Assumption (b) tells us that these accepted solutions are often true. Assumption (a) tells us that the true solution to P is likely to be similar to one of these accepted solutions, which is to say that the true solution is likely to be among those we have considered by this process.

Assumptions (a) and (b) are not unreasonable, and so there is no reason to suppose that Underconsideration is an inevitable problem. Lipton (2004: 162) himself notes "Theory generation is highly constrained by background, and insofar as the background approximates the truth, we should not be so surprised that our powers of generating the truth are substantially better than guesswork." Lipton's reference to 'background' is in the context of background *auxiliary* hypotheses, and it remains unclear how true background auxiliary hypotheses could point our imaginations in the direction of true hypotheses. But if we think of 'background' as referring to our (true) exemplars, then it becomes immediately clear how such constraints are truth-conducive. (Lipton's own answer to Underconsideration is that since ranking at stage 2 is dependent on background theories (auxiliary hypotheses), reliable ranking implies that background theories are often true. So Underconsideration cannot be a problem that generally afflicts our theories. While this argument is valid, the sceptical proponent of the problem of Underconsideration can doubt its premise by happily denying that our stage 2 ranking is known to be reliable.)

I am arguing that if we are in a position that our models are true hypotheses, then modelling our new hypotheses on those will direct our attention to potential explanations that are likely to contain the true explanation among them. Our search in the space of hypotheses is not a random walk. For example, in 1981 researchers found a clustering of highly unusual symptoms, a rare form of pneumonia, *Pneumocystis carinii* (now known as *Pneumocystis jiroveci*), and a rare form of cancer, Kaposi's sarcoma, among homosexual men in the United States. This syndrome, eventually called AIDS, elicited four hypotheses from researchers:

(1) Recreational drugs.
(2) Overload of the immune system from familiar sexually transmitted diseases and other infection.
(3) Bacterial infection—infection by a bacterium, probably hitherto unknown.
(4) Viral infection—infection by a virus, probably hitherto unknown.

These explanations were suggested by researchers' experiences with disease and illness. For example, immunodeficiency was already known to be a consequence of familiar infections, such as chicken pox. Perhaps severe immunodeficiency might be caused by multiple infections, such as young gay men experienced. A natural pair of hypotheses for any new syndrome is viral and bacterial infection, since so many other diseases are caused by viruses and bacteria. And of course in this case the viral hypothesis was shown to be correct by Luc Montagnier in 1983. And so, if by good science or good fortune we have been able to generate a correct explanatory hypothesis in the past and its success leads to it becoming an exemplar that informs future searches for explanations, then there is a high chance that the true answers to those subsequent explanatory puzzles will be generated. In that way, if we ally the exemplar model of the cognitive processes of science with scientific realism, we can see that the former can generate a virtuous cycle of true answers to explanatory puzzles generating further true answers to similar explanatory puzzles.

6 Conclusion

A common conception of cognition says that our reasoning processes, if rational, should be explicable by reference to a priori rules. Rational deductive thought can be explained as conforming to a priori rules of deductive logic, as articulated initially by Peano, Frege and Russell. Carnap and others hoped to extend this to inductive reasoning. Popper, holding that there could be no such logic of induction, maintained that rational inference in science must therefore be deductive (hence his falsificationism). Bayesians claim instead that good scientific reasoning can be understood as instantiating Bayes's rule, itself a theorem of the a priori axioms of probability.

Cognitive scientists and naturalistically inclined philosophers recognise that this conception of rational inference is too narrow. Margolis (1987), for example, argues that pattern recognition is central to human thinking and judgement, including in science. The case made for the role of exemplars in science supports this idea. The human capacity for pattern recognition is a cognitive capacity—it helps us make correct judgements about the world. Yet it cannot be construed as implementing a rule. There is nothing logical (nor illogical) in recognising a face; nor in recognising one scientific problem as like another.

Furthermore, since pattern recognition is a basic cognitive skill, not reducible to the implementation of a rule or the product of some other kind of reasoning, we are typically unable to articulate (from the first-person perspective) what it is in virtue of which we exercise this skill successfully. One often sees a pattern, and as far as the subject is concerned, that is that. In this respect, pattern recognition is phenomenologically like the appreciation of aesthetic qualities: cognitive and aesthetic responses can be elicited immediately, without conscious ratiocination, leading to a conviction that this is correct/beautiful and that can be difficult to justify verbally. (Although neither requires conscious reasoning, both can be enhanced by conscious thought.) These first person similarities explain why scientific cognition using exemplars is often described using aesthetic terminology.

Instances of great art can be held up as exemplars of aesthetic qualities (beautiful, sublime, tragic, romantic, graceful, and so on) and can thereby set the standard by which other works are judged to have these qualities. In the sciences likewise, great achievements—paradigms in Kuhn's terms—exemplify (among other things) our standards of good science. In aesthetics realism is highly contentious, whereas in the philosophy of science realism, although contended, is the majority view. The exemplars of science may be false or they may be true. When true, the standards exemplified by an exemplar will be truth-conducive. If they are false, then we may find false theories more attractive (arguably this was the case for much pre-modern science). It cannot be the *philosopher's* job to say that scientists are now in the former, 'good' scenario of learning our values from true exemplars. And in any case, some fields of science may be in a good scenario and others in a bad scenario. But the philosopher can argue, as I have tried to do, that the good scenario is at least *possible*: learning from exemplars *can* inculcate in us standards of explanatory loveliness that will give us a preference for explanatory theories that are more likely to be true.

Notes

1. See Kuhn (1974) for an account of learning the categories of objects by learning to see similarities and differences, which he thinks is an instance of the same kind of thing as learning to use exemplars in research. Note that Kuhn (1974: 310) remarks that this process can be modelled on a computer; cf. related remarks in Structure (Kuhn 1970: 192).
2. It is notable that Kuhn (1970) cites many experimental psychologists, such as Bruner, Postman, Hastorf, Stratton, Whorf, and the Gestalt psychologists, i.e. Wertheimer, Koffka, and Köhler, in supporting his claims.
3. Duhem's description of good sense ('le bon sens') matches this account of quasi-intuitive capacities in several respects. Good sense gives a scientist those 'motivations which do not proceed from logic and yet direct our choices'. 'Good sense exists in every scientist to some degree . . . and it can be cultivated and sharpened by training and practice', says Ivanova (2010: 60). There are, however, important differences. Duhem does think that good sense can be acquired from exemplars. But the exemplars, according to Duhem, are not instances of great science, but are great scientists (i.e. French scientists plus Newton and Huygens). According to Ivanova good sense is the collection of intellectual virtues possessed by an ideal scientist. This good sense does not vary from field to field or from era to era. Whereas what scientists learn from Kuhnian exemplars is localised to the field from which the exemplars are taken, and will change over time, e.g. as a result of revolutionary change. A final difference that is relevant to my argument (and not one that Kuhn would have endorsed) is that quasi-cognitive intuitive capacities need not be virtuous, in that they need not lead to better theories. Learning with bad exemplars can make one disposed to prefer flawed theories.
4. I do not agree with all the details of Walker's account of loveliness. For example, he regards the 'dormitive virtue' explanation of opium's capacity to induce drowsiness as unlovely (2012: 64). In the context of IBE that would mean that this explanation should be rejected as being unlikely to be true. But it is true, albeit only minimally informative. So although it may in one sense fail to be a 'lovely' explanation, that is not the sense of 'lovely' relevant to IBE.
5. Walker's (2012: 68–69) response to this objection is to argue that subjectivity is really relativity to a scientific community. Exemplars and explanatory goodness do indeed vary between such communities: exemplars and what counts as a good explanation differ between evolutionary biology and geology. But I do not think that this answers the problem—indeed Walker's solution looks like a communitarian version of Lipton's solution. Instead of

focusing on the scientific community *per se*, we should focus on the psychology of individuals that working with exemplars induces. That psychology will indeed be shared by members of the same community, because they work with the same exemplars.

6. I would remark parenthetically that a possible response to Hungerford's objection is just to deny Hungerford's claim. Perhaps there is more objectivity in aesthetic judgement than she suggests, as Aristotle and Hume held, and has been argued by some recent psychologists (see Grammer, Fink, Møller and Thornhill 2003; Dutton 2003; Falk and Balling 2010).
7. Walker (2012: 72) puts it well regarding explanatory loveliness, 'There is no trans-world standard; loveliness is changeable enough to establish and improve a connection with truth in each possible world where IBE is applied'.
8. And I owe a debt to McAllister's (1999) book in developing my thinking on this topic.
9. To give a musical parallel, critics may disagree as to whether a piece is 'beautiful' or 'great' but may agree that it creates a dramatic tension by means of a sustained crescendo over a repeated rhythmic pattern.
10. McAllister does list explanatory power among the empirical criteria of theory assessment.

References

Andersen, H., Barker, P. and Chen, X. (1996), 'Kuhn's Mature Philosophy of Science and Cognitive Psychology', *Philosophical Psychology*, 9: 347–363.

Bird, A. (2005), 'Naturalizing Kuhn', *Proceedings of the Aristotelian Society*, 105: 109–127.

Bird, A. (2007), 'Incommensurability Naturalized', in L. Soler, H. Sankey and P. Hoyningen-Huene (eds.), *Rethinking Scientific Change and Theory Comparison*, Boston Studies in the Philosophy of Science, Vol. 255. Dordrecht: Springer, pp. 21–39.

Bird, A. (2020), 'Scientific Realism and Three Problems for Inference to the Best Explanation', in W. J. Gonzalez (ed.), *New Approaches to Scientific Realism*. De Gruyter.

Carney, D. R., Cuddy, A. J. C. and Yap, A. J. (2010), 'Power Posing: Brief Nonverbal Displays Affect Neuroendocrine Levels and Risk Tolerance', *Psychological Science*, 21: 1363–1368.

Duhem, P. (1914), *La théorie physique, son objet et sa structure* (2nd enlarged ed.). Paris: Chevalier et Rivière. (Reissued, Vrin, 1981) Translated by P. Wiener (1954), *The Aim and Structure of Physical Theory*. Princeton, NJ: Princeton University Press.

Dunbar, K. (1996), 'How Scientists Really Reason', in R. Sternberg and J. Davidson (eds.), *The Nature of Insight*. Cambridge, MA: MIT Press, pp. 365–369.

Dunbar, K. (1999), 'How Scientists Build Models: In Vivo Science as a Window on the Scientific Mind', in L. Magnani, N. J. Nersessian and P. Thagard (eds.), *Model-Based Reasoning in Scientific Discovery*. New York, NY: Kluwer/Plenum, pp. 85–99.

Dutton, D. (2003), 'Aesthetics and Evolutionary Psychology', in J. Levinson (ed.), *The Oxford Handbook of Aesthetics*. New York, NY: Oxford University Press, pp. 693–705.

Falconer, I. (2004), 'Charles Augustin Coulomb and the Fundamental Law of Electrostatics', *Metrologia*, 41: S107–S114.

Falk, J. H. and Balling, J. D. (2010), 'Evolutionary Influence on Human Landscape Preference', *Environment and Behavior*, 42: 479–493.

Gentner, D., Holyoak, K. and Kokinov, B. (2001), *The Analogical Mind: Perspectives From Cognitive Science*. Cambridge, MA: MIT Press.

Glynn, I. (2010), *Elegance in Science*. Oxford: Oxford University Press.

Grammer, K., Fink, B., Møller, A. P. and Thornhill, R. (2003), 'Darwinian Aesthetics: Sexual Selection and the Biology of Beauty', *Biological Reviews*, 78: 385–407.

Heering, P. (1992), 'On Coulomb's Inverse Square Law', *American Journal of Physics*, 60: 988–994.

Heering, P. (1994), 'The Replication of the Torsion Balance Experiment: The Inverse Square Law and Its Refutation by Early 19th-Century German Physicists', in C. Blondel and M. Dörries (eds.), *Restaging Coulomb. Usages, Controverses et Réplications autour de la Balance de Torsion*, Biblioteca di Nuncius, Vol. IV. Florence: Leo S. Olschki, pp. 47–66.

Ivanova, M. (2010), 'Pierre Duhem's Good Sense as a Guide to Theory Choice Good Sense as a Guide to Theory Choice', *Studies in History and Philosophy of Science*, 41: 58–64.

Kihlstrom, J. F. (1987), 'The Cognitive Unconscious', *Science*, 237: 1445–1452.

Kragh, H. (1990), *Dirac: A Scientific Biography*. Cambridge: Cambridge University Press.

Kuhn, T. S. (1970), *The Structure of Scientific Revolutions*, 2nd ed. Chicago, IL: University of Chicago Press.

Kuhn, T. S. (1974), 'Second Thoughts on Paradigms', in F. Suppe (ed.), *The Structure of Scientific Theories*. University of Illinois Press, pp. 459–482. Page references to reprint in Kuhn 1977.

Leake, D. (1998), 'Case-based Reasoning', in W. Bechtel and G. Graham (eds.), *A Companion to Cognitive Science*. Oxford: Blackwell, pp. 465–476.

Lipton, P. (2004), *Inference to the Best Explanation*, 2nd ed. London: Routledge.

Margolis, H. (1987), *Patterns, Thinking, and Cognition: A Theory of Judgment*. Chicago, IL: University of Chicago Press.

Martínez, A. A. (2006), 'Replication of Coulomb's Torsion Balance Experiment', *Archive for History of Exact Sciences*, 60: 517–563.

McAllister, J. (1999), *Beauty and Revolution in Science*. Cornell University Press.

Nickles, T. (2003), 'Normal Science: From Logic to Case-based and Model-based Reasoning', in T. Nickles (ed.), *Thomas Kuhn*. Cambridge: Cambridge University Press, pp. 142–177.

Stanford, P. K. (2006), *Exceeding Our Grasp: Science, History, and the Problem of Unconceived Alternatives*. New York, NY: Oxford University Press.

Strack, F., Martin, L. L. and Stepper, S. (1988), 'Inhibiting and Facilitating Conditions of the Human Smile: A Nonobtrusive Test of the Facial Feedback Hypothesis', *Journal of Personality and Social Psychology*, 54: 768–777.

Walker, D. (2009), *A Kuhnian Defence of Inference to the Best Explanation*. Ph.D. thesis, University of Bristol.

Walker, D. (2012), 'A Kuhnian Defence of Inference to the Best Explanation', *Studies in History and Philosophy of Science*, 43: 64–73.

8 The Aesthetic and Literary Qualities of Scientific Thought Experiments

Alice Murphy

1 Introduction: The Aesthetics of Thought Experiments

The discussion of aesthetic value in science has primarily focused on the evaluation of theories or of mathematical proofs (Breitenbach 2015; Ivanova 2017a, 2017b). Questions then arise regarding the role of these evaluations, such as whether they have an epistemic function, that is, as an indicator of the truth of the theory and/or whether they can aid understanding, as well as the usefulness of such value in science. One aspect of scientific practice that has an obvious aesthetic dimension, but is currently overlooked in the literature, is the use of thought experiments. Thought experiments are a popular device in science and philosophy that are used to justify, undermine or clarify theories. Thought experiments take the form of a description of an imaginary scenario, followed by a judgement of what might happen if the scenario occurred in reality. We then draw conclusions in order to say something about other cases. Examples include Galileo's falling bodies, Thomson's violinist, Maxwell's demon, Schrödinger's cat, brain in a vat scenarios, and Einstein's use of thought experiments in the development of special and general relativity.

Thought experiments are often referred to as beautiful or elegant. Take, for example, Galileo's famous falling bodies thought experiment used to undermine Aristotle's physics, what Brown refers to as 'the most beautiful thought experiment ever devised' (2004: 24).[1] The thought experiment begins by considering Aristotle's theory that heavier bodies fall faster than light ones. Galileo asks us to imagine taking a heavy cannonball, and a lighter musket ball, and dropping them from a tower. According to Aristotle, the heavy ball will fall much faster than the lighter ball. But Galileo points out that if we then imagine attaching the two balls together, and then dropping them, Aristotle's theory leads us to a contradiction. Because it states both: 1) that the compound object will fall faster than the heavy ball on its own, because the compound object is heavier, and 2) that the compound object will fall slower as the lighter ball is inclined to fall slower than the heavy ball, and so it will drag it back—making it fall slower. Both of these predictions cannot be true, and so, Aristotle's theory is contradictory and has to be rejected. From this thought experiment Galileo establishes a new theory that moving bodies fall at the same speed.

In 2012, edge.org conducted a survey that asked 192 people, including scientists and philosophers, what their 'favourite deep, elegant or beautiful explanation' is. As Stuart points out, 21 of the answers given were thought experiments, and a further eight were 'imagination-based inferences that any broad-minded characterization of thought experiments should include'. This means nearly 1/6 of all replies provided a thought experiment as their answer (2018: 530). Physicist Sean Carroll's favourite deep, elegant or beautiful explanation, for instance, is Einstein's thought experiment used as part of his explanation of why gravity is universal, what Einstein called the 'happiest thought' of his life. While Stuart is interested in how this supports the claim that thought experiments provide good explanations and can contribute to scientific understanding, I am interested in the widespread view that thought experiments have aesthetic value.

It is clear, then, that thought experiments are often taken to have aesthetic value. But why are certain thought experiments beautiful or elegant? As Sibley points out in his influential paper on aesthetic concepts, when we describe something using aesthetic terms such as 'unified', 'serene', 'dynamic', 'vivid', 'balanced', 'graceful', or 'elegant' (to take just some of his examples of aesthetic concepts), we often point to non-aesthetic features to explain our application of an aesthetic term. Sibley offers the following examples, 'delicate because of its pastel shades and curving lines', or 'it lacks balance because one of one group of figures is so far off to the left and is so brightly illuminated' (1959: 424).[2] We can think about this in the science case as well, and identify the non-aesthetic features of theories, models, thought experiments and so on, that help explain our application of aesthetic terms. First, what are these features in the case of theories? This is not always clearly set out, and often beauty is described by referring to other aesthetic terms such as simplicity, symmetry or harmony. But to take one example, Poincaré reduces beauty to simplicity and unity. A theory is simple because of the 'number of hypotheses and axioms of the theory. Syntactic elegance or simplicity can be understood as the lack of complexity, adhocness, or free parameters in a theory' (Ivanova 2017b: 2585). The unity of theories is a matter of finding 'hidden relations' between phenomena that appear disconnected (2017b: 2588). A clear feature that is part of the beauty of theories is economy; the theory postulates a small number of hypotheses and axioms, which provide many successful predictions, or can explain a wide range of phenomena.

While most accounts of the aesthetics of science have focused on theories, there has been some consideration of the aesthetics of experiments. This is perhaps a more useful comparison to thought experiments than theories are, as, although the question of whether thought experiments classify as genuine experiments is disputed, they share some important features with ordinary experiments. Unlike theories, both thought experiments and experiments involve (real or imagined) particulars, and there is an initial set up of the experiment, or description of the scenario, which is then manipulated to see or consider what would happen. The main difference is that unlike experiment, thought experiments take place in the imagination or what Brown (2011) calls 'the laboratory of the mind' rather than intervening on the world. Parsons and Rueger have discussed scientists' aesthetic responses to certain experiments. Experiments may be considered beautiful because they produce phenomena that are aesthetically pleasing to experience, for example, Canton's electric *aurora borealis* experiment (2000: 408). Some thought experiments can also be seen as beautiful in this sense, such as Einstein's thought experiment that gets us to imagine the experience of chasing a beam of light.

Parsons and Rueger note that the prevailing eighteenth-century view was that an experiment was beautiful when it 'made visible particular aspects of the beauty of nature itself' (409). Since the nineteenth-century, another way of thinking about the aesthetics of experiments has emerged. Parsons and Rueger show this through the example of Rutherford and the artificial disintegration of atomic nuclei, described by Peter Kapitsa (1968) as a 'most simple experiment', that led to 'striking results'. Here, an understanding of what is being tested becomes central to aesthetic appreciation. This is a move away from admiring the workings of nature, which can be done irrespective of whether or not we have a grasp of the theoretical framework involved. Parsons and Rueger note that since this shift, a common way of characterising aesthetically valuable experiments is to say they involve 'an optimal use of minimal material':

> An experiment now is aesthetically valuable because it shows 'aptness' in relation of result and tools, of plan and success; it is a beautiful artefact, a manifestation of human ingenuity, an instrument optimally suited to achieve its purpose. What is appreciated is, for instance, the simplicity of the arrangement, its economy, or its ability to unify several tasks in one display.
>
> (411–412)

We have already seen this idea of beauty because of 'optimal use of minimal material' in the discussion of theories. It is also present in the aesthetic judgement of thought experiments, where the potential heuristic fertility of thought experiments is emphasised. For Brown, Galileo's thought experiment is beautiful because it is 'brilliantly original and as simple as it is profound' (2004: 24). Similarly, Carroll states 'Einstein, in his genius, realized the profound implication' of the situation described in the thought experiment. In the experiment case, the material is concrete objects. Thought experiments, of course, differ here; economy is achieved through the particulars that we are prescribed to imagine.[3]

To further illuminate these features in the case of thought experiments, it might also help to contrast these with cases of thought experiments that do not fulfil their function, and could be regarded as cluttered or clumsy and so on. Norton describes Szilard's version of Maxwell's demon as 'the worst thought experiment'. Why is this? Norton offers a detailed account of the thought experiment and its flaws, but we can focus on a couple of reasons he provides. Thought experiments are illuminating when they provide us with a simple scenario that allows us to focus on the most essential features, and when the scenario can serve as a representative case. Szilard's thought experiment involves a misuse of idealisations; 'an inconsistent muddle of improper idealizations', and leads to an incorrect generalisation (2018: 466). Another example is Darwin's whale thought experiment, which attempts to explain natural selection by demonstrating how whales could have evolved from bears. This was described by Louis Agassiz as 'truly monstrous', and was dropped by Darwin in later editions of *On the Origin of Species*. In attempting to explain the morphology of whales by referring to an existing creature, the example invokes 'needlessly strange evolutionary explanations' and fails to aid our understanding of natural selection (Stuart 2016: 31).

Now, it could be argued that when a thought experiment is described as beautiful, or when Darwin's whale is described as monstrous, what is really being said is that the thought experiment is successful or unsuccessful.[4] This worry has been raised in the context of the aesthetics of theories by Todd who claims that 'there are strong grounds for suspecting that what appears to be aesthetic claims may often be, if perhaps not always are, really masked "epistemic" functional ones' (2008: 77). Any account of aesthetics in science will have to provide reasons as to why we should take these descriptions as genuinely aesthetic (for discussion and response, see O'Loughlin and McCallum 2019). Another concern is beautiful theories that turn out to be false, or ugly theories (or at least, ones that are not considered to have aesthetic merit) that are successful, such as the standard model of particle physics (Ivanova 2017a). There may be problems, then, with regards to the epistemic value of aesthetic qualities when it comes to theories. There is clearly more to be said here, but I want to suggest that the function of a thought experiment is closely tied to aesthetic features.[5] The value of thought experiments is that they provide us with a scenario that makes something complex easy to visualise and to grasp. And so, the aesthetic virtues of thought experiments are significant in that a thought experiment that is cluttered or clumsy would be less useful or insightful due to the role that thought experiments play in science, in particular, their role in understanding and pedagogy.

The beauty of thought experiments appears to lie in their ability to evaluate, explain or help us understand something profound based on a simple scenario. Furthermore, beautiful or elegant thought experiments are those that have carefully selected particulars. This leads us to the two ways we can think about the aesthetic dimension of thought experiments. The first, a narrow sense, is to do with their assessment as 'beautiful' or 'elegant' as presented above. In this sense, as we have seen, an aesthetic evaluation of thought experiments may be similar to that of theory and experiment with regards to the 'optimal use of minimal material', that is, simplicity combined with significant consequences. The second, a broader sense, also includes the features that thought experiments have in common with artworks. The rest of the chapter takes the aesthetics of thought experiments in this second, broader sense. In particular, I look to the similarities thought experiments share with works of literary fiction. Despite what has been indicated so far, this second way highlights the ways in which thought experiments are distinct from experiments and theories with regard to their aesthetic qualities.

2 Literature as Thought Experiment

Thought experiments can be characterised as taking the form of short, fictional narratives that have the purpose of instructing the reader to evaluate the described scenario in a certain way. In the philosophy of art, comparisons have been drawn between thought experiments and artworks, particularly works of literary fiction, as they share (at least some of) the key features of thought experiments, namely their fictionality—the events have not actually taken place, or at least this is inessential—and narrative form.[6] Further, the use of thought experiments in learning has been offered as a way of defending the cognitive value of literature. We can, of course, learn about art from engaging with artworks; reading Kafka's *The Metamorphosis* teaches us something about the novel and about Kafka's literary style. Similarly, through literature we could learn about historical, geographical or scientific truths. The interesting issue is whether artworks can teach us in a way that is less constrained than this. Describing fiction as a kind of extended, more complex thought experiment allows us to maintain that engaging with narrative art can lead us to new insights about the world and ourselves. For example, Carroll argues that Graham Greene's *The Third Man* is a thought experiment that presents a powerful counterexample to the maxim: 'When loyalty to a friend conflicts with loyalty to a cause, one ought to choose in favour of the friend' (2002: 10). And for John, fictions such as Paley's short story *Wants* function like a philosophical thought experiment 'in which problematic imagined cases are used to prompt responses relevant to philosophical problems'. John argues that in our engagement with *Wants*, we are led to explore the concept of desire, which is similar to the role of thought experiments in addressing questions about our conceptual schemes (1998: 332).

While many discussing thought experiments in the philosophy of literature have focused on examples in philosophy, Elgin (2014) and Davies (2007) have used thought experiments as a way of bringing together issues in philosophy of art and philosophy of science. It is scientific thought experiments that are my focus here. Let's look at another example, this time from Newton. The thought experiment sets out to undermine Descartes' relational account of motion. We are asked to imagine a bucket hanging from a long rope that is twisted tight. The bucket is then filled with water and the rope is released, making it unwind and the bucket spin. At first, the water and the bucket are in relative motion, and the water is still flat in the bucket. But after some time, the water will pick up the motion of the bucket, forming a concave shape as the water rises up the sides of the bucket. Now there is no relative motion between the water and the bucket—which is how it was in the initial starting point, before the rope was released. So how do we explain the observed difference between the first state (before the bucket was released) and the final state (when the water forms a concave shape), when in each case, there is no relative motion? Newton explains that the motion of the water is absolute and not relative to the bucket, and has to be represented as such in absolute space (Brown 2011: 8).

Elgin argues there is a continuity between concrete experiment, thought experiment, and literary fiction. Experiments and thought experiments involve studying an object or a system that stands in for a target system, and they each require us to control our (real or imagined) set up, ensuring that we carefully isolate the features that we are interested in investigating. We can note here a key difference between the Galileo example presented in the last section and Newton's bucket thought experiment, that demonstrates how much scientific thought experiments can vary with regards to their departure from how things are. As Elgin notes, like ordinary experiment, thought experiments involve the study of simplified and distorted versions of nature. In the Galileo case, there is a more straightforward idealisation; the abstraction of air resistance. Further, we can easily imagine ourselves going to the top of a tower and performing the thought experiment. Newton is taken to be describing an actual experiment in *The Principia*, but in order for the theoretical conclusion regarding the existence of absolute space to follow, there are more demands on our imagination. We are required to imagine that there is nothing in the universe except the bucket filled with water hanging from a rope. Although the rope remains, it is not tied to anything, and even the earth (whose gravity keeps the water in the bucket) does not exist.

For Elgin, this control of our scenario, and the use of idealisation carries over to literary fiction: 'a work of fiction selects and isolates, contriving situations and manipulating circumstances so that patterns and properties stand out'. On this view, fiction functions as a thought experiment that provides us with new insights or understanding, it 'may frame or isolate mundane features of experience so that their significance is evident. It may defamiliarize the commonplace, making us aware of how remarkable normal behavior can be' (2014: 232). As with thought experiments, works of literature differ in terms of how much, and in what way, they depart from reality.

And so, there are views that maintain that (some) narrative art functions like thought experiments, and that this helps explain how we can learn from engaging with these works. Consequently, an analysis of thought experiments and their epistemological value has offered a way of drawing fruitful analogies between philosophy of science and philosophy of fiction. Here, I put to one side the question of whether literary works can be thought experiments, and instead want to think about how the comparisons can be drawn the other way as well, that is, from aesthetic and philosophy of art in order to illuminate the science cases. My aim is to address how the aesthetic choices scientists make in the design of thought experiments contribute to the function of the thought experiment: to communicate, convince, or explain a theory or phenomena to a scientific or a public community. I argue that when thinking about the commonalities between scientific and artistic representations, thought experiments are a fruitful case study for philosophers of science. Part of their value in science is the features they share with literary works. The key issue is whether the aesthetic qualities provide anything beyond catching and maintaining our attention or at best, are a mere heuristic aid. Now I'm going to consider accounts that argue this way. This set of views claim that there are disanalogies between the art and science cases that undermine purported connections between how we learn from scientific and artistic representations, and the role of aesthetic considerations in science. In the next section, I consider two disanalogies between literature and thought experiments. The first is to do with the importance of the formulation of literary works, and the second is to do with the ways in which we interpret artistic fictions on one hand, and literary works on the other.

3 Disanalogies Between Literature and Thought Experiment

3.1 Formulation

Any account that discusses the aesthetic or literary qualities of thought experiments is going to face opposition from Norton. Norton analyses thought experiments as arguments, insisting that they work by transforming our existing knowledge through a logical process. He maintains that all thought experiments can be reconstructed into argument form without any epistemic loss, and that 'the actual conduct of a thought experiment consists in the execution of an argument' (2004: 50). For Norton, this view is the only alternative to the claim that thought experiments, in virtue of providing us with knowledge about the world without new empirical data, are cases of 'epistemic magic'. Norton has in mind Brown's platonist account, whereby some thought experiments work by giving us access to the laws of nature that exist in a platonic realm (Brown 2011, 2004). Norton has reconstructed many thought experiments into arguments, and holds that there are no examples that cannot be handled in such a way. Consequently, their typical narrative form and any of their aesthetic qualities are irrelevant to the conclusion and therefore dispensable.

Egan's view is less strict than Norton's in that he is not committed to the claim that thought experiments are arguments, or that the actual conduct of a thought experiment consists in the execution of an argument. But for Egan, there is an essential connection between scientific and philosophical thought experiments and argument that leads to a crucial disanalogy between thought experiments and literary works. Thought experiments are always a part of a larger, argumentative structure.[7] Otherwise, Egan states, they would be merely 'intriguing narratives' (2016: 142). Egan allows that a work of literature could be used as a thought experiment, that is, a philosopher could use a work of literary fiction or sections of those works as part of a thought experiment or an argument. The same excerpt can function differently in being put to work in this way, for example, a work of philosophy that argues in favour of the importance of freedom in our choices, even when this may lead to us making morally reprehensible decisions, might cite an excerpt from Burgess's *A Clockwork Orange*. However, in a case like this, Egan stresses that the work is no longer being read as literature. Instead, we read it with a particular purpose in mind, that is, as part of an argument for a philosophical position (2016: 143).

Further to this, Egan claims that the purpose of a thought experiment 'is *exhausted* in making or contributing to an argument'. Like Norton, Egan maintains that any aesthetic virtues of a thought experiment narrative are irrelevant to the purpose of a thought experiment; 'we can, as it were, throw away our thought experimental narrative once its work is done' (2016: 142). Because of this, Egan claims that 'their role in arguments is also fungible'. In the case of, say, Nozick's experience machine, a different thought experiment could have been used in its place if it did the same argumentative work—nothing in the argument against hedonism requires the story of the experience machine, whereas we do not approach literature in this way. We do not have a sole purpose in mind for an artwork, such as making an argument. Rather, there is a plurality of reasons for engaging with literary fiction. This means that literary works are not replaceable in the way thought experiments are. Egan gives the example of deriving amusement from a literary work. We do not regard the work simply as a way of deriving amusement, nor do we think that any other equally amusing story could be in its place—he states, 'the uniqueness of just *this* story remains' (2016: 143).

For Egan, a key part of this is that unlike literary works, thought experiments are not concerned with particulars. When presenting the experience machine thought experiment, Nozick (1974) is not concerned with the idea of floating in an experience machine with electrodes attached to our heads, nor is Galileo interested in musket balls and cannonballs. These particulars are a means to an end: they serve as a way of exploring more abstract problems. The same cannot be said for works of literature, where 'the concrete particularities of narratives are irreducible parts of what we attend to when we read a narrative as literature'. Literary works can of course prompt more abstract or general reflections, but these (compared to thought experiments) have not got a defined limit and so, we cannot abstract way from the 'concrete particularities of [for example] Kafka's narrative if we want to think about it as a work of literature' (2016: 144). In addition, Egan states that literature does not have to prompt this type of reflection, 'I can derive great interest and pleasure from reading *Pride and Prejudice* as a story about Elizabeth Bennett' and so on, without drawing more general conclusions about say, class or marriage (2016: 145).

The issue here appears to be about what Stecker (2010) calls the unique value of artworks, and the indispensability and irreplaceability in the case of literary works, which Egan contrasts with thought experiments. The aesthetic experience produced by say Kafka's *Metamorphosis* could not be produced by anything else, and so it cannot be replaced by another work of art of comparable worth. Further, the work is not dispensable once we have read it, that is, it is not a means to an end. Rather, we come back to artworks with the expectation that we can gain something new from them and that they can offer new experiences. Consequently, they offer new value each time.

For Egan and Norton, then, the role of narrative, aesthetic qualities and use of particulars merely make a thought experiment more interesting or compelling. As Norton summarises, 'it is merely rhetorical window dressing that, for psychological reasons, may well ease acceptance of the result. In many cases, this superfluity is easy to see, since the elements visualised can be supplied in many ways that will not affect the outcome' (2004: 60).[8]

3.2 Interpretation

The first disanalogy between thought experiments and literary works emphasised the importance of the format of artistic representations. A second, related disanalogy that has been raised focuses on the ways in which we interpret artworks such as literature on one hand, and scientific representations on the other.

Frigg and Nguyen (2017) highlight the similarities between representation in art and science, and they apply their own framework—the DEKI account, where representation consists in denotation, exemplification, keying-up and imputation—to both scientific and artistic cases. However, they outline an important difference, which is to do with what they call 'the flexibility of interpretation' in artistic representations compared to scientific ones. In the case of models, they claim that the interpretation 'is usually fixed by the context and the interpretation highly regimented. Someone who doesn't interpret the large sphere as the sun simply doesn't understand the Newtonian model' of the solar system (2017: 57). In works of literature, the interpretation is not fixed and attending carefully to the work and its features in order to come up with interesting and sometimes conflicting interpretations is part of engaging with and appreciating artistic works.

Similarly, when it comes to thought experiments, Hacking argues that in contrast with experiments, thought experiments do not have a life of their own. He states: 'I think of [concrete] experiments as having a life: maturing, evolving, adapting, being not only recycled but also, quite literally being retooled. But thought experiments are fixed, largely immutable' (1992: 307). By this he means that thought experiments do not evolve—they cannot be adapted or reworked for different purposes. For Hacking, this feeds into the epistemic privilege of experiment over thought experiment. In the next section, I address these arguments, and argue for the importance of the literary qualities of thought experiments.

4 The Literary Qualities of Thought Experiments

I agree that there are significant differences between our engagement with art on one hand, and our engagement with science on the other, and these differences need to be taken into consideration when drawing comparisons between scientific and artistic representations and how we learn from them. But I want to resist the force of the above claims. I begin by addressing the importance of formulation in thought experiments before demonstrating how thought experiments can be interpreted in different ways.

4.1 Formulation and Concrete Particulars

As seen, Norton states that thought experiments can be reconstructed as arguments without any epistemic loss. I argue that while we can, of course, rationally reconstruct thought experiments into argument form, this will lead us to miss important features involved in their practice, and this is what I am interested in examining. What appears to be implicit in Norton and Egan's view is a distinction between the context of discovery and the context of justification. The idea would be that the narrative form of a thought experiment and their use of particular elements (the use of ordinary or familiar objects in thought experiments, for example, Newton using buckets and ropes to show the existence of absolute space) may play a role in the context of discovery, that is, in the conduct of a thought experiment, but it is limited to this—they have no part to play in the context of justification.

There are two ways of thinking about this distinction in the case of thought experiments: the first is to do with what went on in the mind of the scientist who initially conducted the thought experiment, which brought them to a new insight. Take, for instance, Einstein's recollection in his autobiography, of him imagining chasing after a beam of light when he was 16, which was an important part of the development of special relativity as mentioned in footnote 6. We can label this a private thought experiment. A second way to think about thought experiments is at the stage in which they have been presented to a public or a scientific community in order to communicate an idea—such as in the Galileo and Newton cases presented above. In these cases, the context of discovery can be taken to embrace the cognitive processes of agents when performing a thought experiment, and what the thought experiment asks us to do. It is the latter that I focus on.

The discovery/justification distinction has received a lot of attention, and many have undermined attempts to draw a clear-cut distinction or have argued that the context of discovery is a legitimate topic for philosophers of science (for an overview see Schickore 2018). The distinction appears especially difficult to apply in the case of thought experiments, as it is in the performance of a thought experiment that they get their justificatory force. Against Norton's reconstruction thesis—that thought experiments can be reconstructed as arguments without epistemic loss—Gendler has shown that the demonstrative force of the thought experiment depends upon its formulation, that is, its narrative form and reference to concrete particulars, making these features indispensable. We can see this in the case of Galileo's falling bodies thought experiment. There are various logical 'ways out' for the Aristotelian (who holds that heavier bodies fall faster than lighter ones) that are not available when the thought experiment is presented in its original, unreconstructed form. For example, considering the difference between the two bodies attached together as a new, unified object, compared to seeing it as a composite object made of two objects. Gendler states in the argument version, we lose 'the way in which, by evoking tacit knowledge about how the falling bodies actually behave, the thought experiment pre-emptively precludes such ways out' (1998: 407).

We can see the importance of formulation in other scientific representations, such as scientific models. Frigg and Nguyen (2017) argue that the 'very same model, when presented under a different format, can yield different predictions and offer different explanations. Formulation matters' (2017: 58). They cite the work of Vorms who argues that different formats of the same information impacts agents' reasoning processes, and that different formats allow access to different information. One case

Vorms discusses is representing the results of a temperature survey in different ways. While it is the same information in each representation, 'the map makes some information much more easily available: for instance, if warm shades stand for high temperatures and cold shades for low temperatures, one can quickly conclude that the southern part of the represented area is warmer than its northern part'. Whereas drawing this from a different representation of the same information, say a list of numerals, would involve many inferential steps (2011: 289).

Given this focus on the practical features of representing in the scientific context, it is difficult to see the force of the claim that the particulars do not matter in the case of thought experiments, and that they can be dispensed with. It is through the introduction of the particulars and ordinary and/or familiar objects—Newton using buckets and ropes, Galileo towers and balls and so on—that we engage with the thought experiment, and therefore come to understand what the thought experiment and the relevant theorising is about. The particularities are important and carefully chosen. Thought experiments are devices of the imagination, and depend upon our imaginative capacities. The use of objects gives our imagination something to latch onto, to help us work through the problem. In Galileo's case, it is the two balls attached together and imagining their behaviour that is an integral part of the presentation.[9] Furthermore, if we look at the context in which *The Two Dialogues* was written, Galileo was writing to appeal to not only a scientific, but a public community. The particulars used are suited to that audience, and enhance the accessibility of the thought experiment scenario, consequently contributing to the cognitive force of the thought experiment.

Egan could still maintain that although the role of concrete elements in thought experiments is indispensable in some sense, this does not fully address his concern. As we have seen, Egan claims that in the case of literature, 'the concrete elements of the narrative remain irreducibly a part of our imaginative engagement' (2016: 144). They are ineliminable and irreplaceable in that they are not a means to an end—treating literature as thought experiment suggests that, for example Tolstoy's *The Death of Ivan Ilyich* 'exhausts its purpose once it has made a particular distinction salient, and we could just as well used some other thought experiment provided it made the same distinction equally salient' but when treating it as a work of literature, it is not exhausted or replaceable in this way (2016: 143). It seems true that there is a particular experience of reading a work of Tolstoy that cannot be had another way, and that the details of a thought experiment could be changed to some degree without impacting its effectiveness, but there are a couple of worries here. First, it appears difficult to identify exactly what changes in the features of a thought experiment are permissible without altering the thought experiment, or its appeal and force and therefore usefulness in scientific practice. Further, it appears that at least some of the details of a work by, say,

Tolstoy, could be changed without altering the novel, or affecting the interpretations we draw from the novel.

Second, we can ask whether all artworks, uniquely valuable in this way, are irreplaceable and indispensable? Stecker is critical of the irreplaceability claim, he states that this 'is much less likely to be true for a certain fishing rod model, but it is also not true for many lesser works. In both cases, these sorts of things are constantly going out of existence or becoming unavailable without a great loss of value in the world' (2010: 89). It seems plausible that some artworks could be replaced, or that some elements of artworks could be replaced, without a change in aesthetic value and/or experience, and so a sharp divide between thought experiments and literary works cannot be drawn by virtue of this, for example, generic, formulaic pieces of music or novels, or certain forms of kitsch art. In section 5, I suggest we ought to look closely at the types of literary fictions used when drawing analogies and disanalogies with thought experiments.

4.2 Flexibility of Interpretation

The second issue I will offer a response to is to do with the flexibility of interpretation in artistic and scientific representations as raised by Frigg and Nguyen. First, we can address this on the philosophy of art side. Frigg and Nguyen's example of interpretation in the scientific context is highly regimented, someone 'who doesn't interpret the large sphere as the sun simply doesn't understand the Newtonian model' is true, but this seems too simple and has obvious parallels in the interpretation of artworks. If I don't take the *Mona Lisa* to be a representation of a woman, or at least of a person, then I have clearly misunderstood the painting. Here, it might help to appeal to a distinction between a description of a work (or representation more generally), and an interpretation of that work (Matthews 1977). This distinction can apply to both the art and the science cases. That the largest sphere is the sun is part of the description of the model, similarly, that the *Mona Lisa* is a portrait is part of the description of the painting. This differs from offering an interpretation of the representation. In addition, there is huge debate in the philosophy of art regarding the interpretation of artworks, and whether there are several correct or acceptable interpretations of a work (pluralism), or that there is a single correct or acceptable interpretation (monism). Even if it is allowed that there can be multiple, inconsistent interpretations of a work, it is not the case that 'anything goes' when interpreting works of art (for discussion, see Stecker 1994).

We can also address this on the philosophy of science side, and show that thought experiments are not limited to a single interpretation. As Bokulich shows with the case of the Einstein-Podolsky-Rosen thought experiment, which asked whether quantum mechanics is complete, 'thought experiments are no more bound to any one particular theory than ordinary physical experiments are' (2001: 293). There can be disagreement regarding what exactly a thought experiment shows and it can be analysed from the perspective of different theories, in this case, quantum mechanics and hidden variable theories. The same goes for the Newton's bucket outlined in section 2. Mach presents an alternative analysis of the thought experiment, stating that it 'does not establish the existence of absolute space or absolute motion, only that motion relative to the Earth or fixed stars produces such effects (whereas the water's motion relative to the bucket does not)' (Bokulich and Frappier 2017). This demonstrates his questioning of Newton's omission of certain particulars that Mach took to be crucial—namely the fixed stars, which provides the relata within his Leibnizian conception of space. Further, some thought experiments are still being revised and in some cases, their impact remains a contentious issue. An example is Einstein's clock in the box thought experiment that sets out to falsify Heinsenberg's uncertainty principle. It remains unclear how the thought experiment will be resolved; it 'continues to be debated in the same context as it was originally presented, namely, concerning whether and how the uncertainty principle maintains itself in the face of certain possible experimental arrangements', and different formulations are still provided (Stuart 2016: 28).

While the interpretation of thought experiments has limits, it is underdetermined—there can be disagreement on what would happen in the scenario presented, or what conclusion should be drawn. And this is part of thought experiments' value in science. Further, considering alternative interpretations of what phenomena would occur in a thought experiment setup, or what the thought experiment actually demonstrates can also be a part of the style of presenting thought experiments. We see this in Galileo's dialogues where differences amongst the interlocutors' interpretations are presented, and these interpretations depend on their different theoretical commitments (Palmerino 2018).

5 Selecting the Right Examples

I want to end by suggesting that we ought to look closely at the literary examples used when thinking about the qualities that thought experiments share with works of literature. If we have in mind the likes of Tolstoy, then similarities are going to be thin. Further, such works are not representative of all narrative art. Perhaps some short stories would be a better comparison, particularly those that are revolved around a certain theme or a clear idea. One example is the 2010 short story collection *Machine of Death* where each story centres on the exploration of the same premise; a device that identifies how a person will die based on an analysis of a blood sample.[10] Further, Cameron's discussion of speculative fiction as moral and metaphysical thought experiment shows that works of literary fiction vary with regards to their similarities with thought experiments. Unlike works of realist fiction, speculative fiction as a genre describes worlds that depart radically from our own, and are not constrained by the history of our world. In creating such 'extreme worlds', relevant features can be isolated and exaggerated. Cameron gives the example of Orwell's *1984*, and compares it with works of realist fiction that also depict totalitarianism:

> Speculative fiction can *aim at*, and to varying degrees be successful at, distilling the essence of what it is to be focused on in a way that realistic fiction by its very nature cannot, since the latter must inevitably not focus on totalitarianism *per se* but rather totalitarianism *within* the broader social-historical context that is the setting of the novel.
>
> (Cameron 2015: 33)

And so, we should be careful when selecting our points of comparisons when discussing the aesthetic qualities of thought experiments; some literary fictions will be more relevant than others. One form of short literary work that provides a useful comparison is fables and parables. Take an example of a fable from Aesop (from Hunt 2009: 370):

> Between the North Wind and the Sun, they say, a contest of this sort arose, to wit, which of the two would strip the goatskin from a farmer plodding on his way. The North Wind first began to blow as he does when he blows from Thrace, thinking by sheer force to rob the wearer of his cloak. And yet no more on that account did he, the man, relax his hold; instead he shivered, drew the borders of his garment tight about him every way, and rested with his back against a spur of rock. Then the Sun peeped forth, welcome at first, bringing the man relief from the cold, raw wind. Next, changing, he turned the heat on more, and suddenly the farmer felt too hot and of his own accord threw off the cloak, and so was stripped.
>
> Thus was the North Wind beaten in the contest. And the story means: 'Cultivate gentleness, my son; you will get results oftener by persuasion than by the use of force',

As with literary works and thought experiments, fables and parables such as this focus on examples of concrete objects and events. Like thought experiments, they are not merely a short fictional story, nor are they simply an articulation of a viewpoint, but are written with a purpose to persuade or explain something to the reader. In this case, the fable's moral is that persuasion is superior to force. Further, the characters and objects in a fable are idealised and depart from their real world counterparts. The situations are simple, they are not situated in a certain historical context or geographical location, and the conclusion is intended to be generalised in order to apply to a broader range of cases than depicted in the example. At the same time, although their relation to literature is contentious, their style is clearly literary rather than that of abstract argumentation.

Given these similarities, we can think about the connection between thought experiments, and fables and parables through genre. Genres are categories of artworks, and a work being in a particular genre influences the way in which it is experienced, understood and evaluated by the reader or audience (Friend 2012). The idea of thought experiments as a genre has been explored by Weinberg:

> There are rules to engaging properly with a hypothetical scenario, after all. To make just some of the more obvious generalizations about our imaginative practices with thought experiments: one should embellish as little as possible; generally it is a practice conducted in an affectively 'cool' manner; and our inferential systems must often be brought to bear in this particular sort of imaginative project as well. And there are surely other, and more subtly articulable, rules for the proper performance of thought experiments still to be detailed.
>
> (Weinberg 2008: 214)

In section 1, we saw through Norton's discussion of Szilard's thought experiment, 'the worst thought experiment', that there are certain conventions involved in the creation and engagement of thought experiments with regards to idealisations and generalisability. Moreover, when teaching students how to engage and criticise thought experiments, there is often some work to be done in order to get them familiar with the genre, such as explaining how thought experiments work, why they are used, and what are the right and wrong questions to ask about the imagined scenario. Given the similarities presented, we could hold that thought experiments are a sub-genre of a larger genre that also includes fables and parables. In each, we attempt to take on board the things that are stipulated by the author, and do not object to them on the grounds that they are 'unrealistic', say. There is a shared convention to how we engage with these stories, they have a point or moral, and the intended point guides the reader's interpretation of the story.[11]

In the scientific context, Cartwright (1991, 2010) has drawn parallels between models on one hand, and fables and parables on the other: 'Fables transform the abstract into the concrete, and in so doing, I claim, they function like models in physics . . . the relationship between the moral and the fable is like that between a scientific law and a model' (1991: 57). As Cartwright explains, in order to find a conclusion that is true in both the model or fable and in other cases beyond the particulars in described scenario, we may have to 'climb up the ladder of abstraction', that is, express the conclusion or the moral in more abstract terms. The relation between moral and the fable (or a model/thought experiment and its conclusion) is 'that of the general to the more specific' and the moral is 'fitted out' by the fable. That is, 'the moral describes just what happens in the fable; but the fable fits it out in a special way—a way true to the moral but not necessarily shared by all cases of which the moral is true' (2010: 26–27). In another case, the 'fitting out' of the abstract moral will be very different from say, the sun and the wind in Aesop's fable above, or falling bodies from a tower in the Galileo thought experiment.

A key difference between parables, such as the good Samaritan, and fables such as Aesop's fables, that Cartwright highlights in her later work, is that the former do not typically have the moral or lesson 'built in'. Instead, defending a view of what the parable shows involves interpretative work, including attending to other parts of the text in which it is presented, as well as how the world operates. Cartwright argues that many of the highly idealised models utilised in physics and economics are more like parables than fables in this sense: 'A variety of morals can be attributed to the models, expressed in a variety of different vocabularies involving abstractions of different kinds and at different levels. Importantly, these morals can point in different directions, implying opposite predictions for the real-life situations to which we want to apply them' (2010: 21). In the previous section, we saw how the interpretation of thought experiments is flexible, that theoretical commitments alter the conclusions drawn, and that there is debate around what exactly a thought experiment shows. In Cartwright's terms, the moral of a thought experiment or parable is not part of the parable itself; outside work needs to be done to show which 'ladder of abstraction' must be climbed to reach the result that can be generalised.

6 Conclusion

To sum up, we have seen that some have drawn comparisons between thought experiments in science and philosophy as a way of defending the cognitive value of literature. Here, thought experiments have been utilised to explore connections between scientific and artistic representations in a different way: thought experiments in science are a good case study for thinking about aesthetic features in the scientific context. I hope to have shown that when thinking about thought experiments in science, the difference between representations in art and science is not as stark as some have made it out to be, and that science is more heterogeneous than has been allowed. The differences between scientific and artistic representations raised in current discussions fail to adequately account for the use of thought experiments in scientific practice, and part of their value in this context includes the qualities that they share with literary works.

Acknowledgements

I am grateful to Steven French, Aaron Meskin, Milena Ivanova, and those who attended the White Rose Aesthetics Forum in Leeds, January 2019, for helpful feedback and conversations on earlier versions of this chapter. This work was supported by the AHRC through the White Rose College of the Arts & Humanities.

Notes

1. Brown also cites the results of a poll conducted by *Physics World* of the ten most beautiful experiments of all time. Galileo's thought experiment is second on the list (Crease 2002).
2. It is important to note that Sibley argues that aesthetic terms cannot be defined in terms of non-aesthetic concepts. Todd suggests that whether or not Sibley is correct, it seems that the conditions under which an aesthetic term is employed are clearer in the case of science and mathematics than in art (2008: 71).
3. Bringing thought experiments into the aesthetics of science discussion leads to some further considerations. So far I have presented reasons to suggest that the aesthetic 'canons' we form for theories, experiments, and thought experiments will be the same in that each generate a positive aesthetic response if they are elegant and simple. But there may be reasons to argue that a different set of aesthetic properties could be valued in the three contexts. Further, we can ask whether the force of the aesthetic qualities of thought experiments change from era to era, throughout the history of science. On McAllister's (1996) view, in theory change our aesthetic canons and consequently what counts as beautiful will be revised. Finally, I expect there to be variation with regards to the force of aesthetic qualities of a thought experiment according to the role it is playing in science.
4. I outline Norton's view that thought experiments are arguments below. Given his account, it appears Norton would maintain that any aesthetic qualities of thought experiments are irrelevant to their epistemology, or that the aesthetic qualities can be reduced to claims regarding their role in contributing to an argument and consequently, they are not genuinely aesthetic.
5. This may also apply to concrete experiments in that a complex experiment offers more possibilities for being contested. The aesthetics of science may offer a further way to explore connections between thought experiments and ordinary, concrete experiments.
6. A further issue is whether all scientific thought experiments count as narratives. Here, I am just going to focus on cases that do take a narrative form.
7. A counterexample to this is Einstein's recollection in his autobiography of imagining himself chasing a beam of light, and considering what he would observe. This thought experiment, in its initial stage, was exploratory and separate from any larger argumentative structure.
8. These thoughts are echoed in Currie's discussion of the comparison between scientific models and literary fiction: 'Models are not dependent for their value in learning on any particular formulation; rather they depend on their capacity to get good predictive or explanatory results or to achieve some other epistemic aims' (2016: 305).
9. Similarly, for Einstein's 1905 On the Electrodynamics of Moving Bodies—the theory is formulated in terms of the behaviour of clocks and rods; this is what we are invited to imagine (French 2018).

10. Thanks to Jamie Cawthra for directing me to this example.
11. Weatherson (2010) has also discussed this connection between fables, thought experiments and genre on his blog, under 'surveys and thought experiments'.

Bibliography

Bokulich, A. (2001), 'Rethinking Thought Experiments', *Perspectives on Science*, 9: 285–307.

Bokulich, A. and Frappier, M. (2017), 'On the Identity of Thought Experiments: Thought Experiments Rethought', in M. T. Stuart, Y. J. H. Fehige and J. R. Brown (eds.), *The Routledge Companion to Thought Experiments*. Routledge.

Breitenbach, A. (2015), 'Beauty in Proofs: Kant on Aesthetics in Mathematics', *European Journal of Philosophy*, 23: 955–977.

Brown, J. R. (2004), 'Why Thought Experiments Transcend Experience', in *Contemporary Debates in Philosophy of Science*. Blackwell.

Brown, J. R. (2011), *The Laboratory of the Mind: Thought Experiments in the Natural Sciences*. Routledge.

Budd, M. (1995), *Values of Art: Pictures, Poetry Music*. London: The Penguin Press.

Cameron, R. P. (2015), 'Improve Your Thought Experiments Overnight With Speculative Fiction!', *Midwest Studies in Philosophy*, 39: 29–45.

Carroll, N. (2002), 'The Wheel of Virtue: Art, Literature, and Moral Knowledge', *The Journal of Aesthetics and Art Criticism*, 60: 3–26.

Cartwright, N. (1991), 'Fables and Models', *Aristotelian Society Supplementary Volume*, 65: 55–82.

Cartwright, N. (2010), 'Models: Parables v Fables', in R. Frigg and M. Hunter (eds.), *Beyond Mimesis and Convention—Representation in Art and Science*. Boston Studies in the Philosophy of Science. Springer.

Crease, R. (2002), 'The Most Beautiful Experiment', *Physics World*, May 02.

Currie, G. (2016), 'Models as Fictions, Fictions as Models', *The Monist*, 99: 296–310.

Davies, D. (2007), 'Thought Experiments and Fictional Narratives', *Croatian Journal of Philosophy*, 7: 29–45.

Davies, S. (2006), 'Authors' Intentions, Literary Interpretation, and Literary Value', *The British Journal of Aesthetics*, 46: 223–247.

'Edge.org', Accessed March 10, 2019 online at www.edge.org/responses/what-is-your-favorite-deep-elegant-or-beautiful-explanation.

Egan, D. (2016), 'Literature and Thought Experiments', *The Journal of Aesthetics and Art Criticism*, 74: 139–150.

Elgin, C. Z. (2014), 'Fiction as Thought Experiment', *Perspectives on Science*, 22: 221–241.

French, S. (2018), 'Imagination in Art and Science', Presentation as part of *Imagination in Science* Symposium, PSA Biannual Meeting, Seattle.

Friend, S. (2012), 'Fiction as a Genre', *Proceedings of the Aristotelian Society*, 112: 179–209.

Frigg, R. and Nguyen, J. (2017), 'Of Barrels and Pipes: Representation—As in Art and Science', in O. Bueno, G. Darby, S. French and D. Rickles (eds.), *Thinking About Science and Reflecting on Art: Bringing Aesthetics and the Philosophy of Science Together*. Routledge, pp. 41–61.

Gendler, T. S. (1998), 'Galileo and the Indispensability of Scientific Thought Experiment', *The British Journal for the Philosophy of Science*, 49: 397–424.

Gendler, T. S. (2010), *Intuition, Imagination, and Philosophical Methodology*. Oxford University Press.

Hacking, I. (1992), 'Do Thought Experiments Have a Life of Their Own? Comments on James Brown, Nancy Nersessian and David Gooding', *PSA: Proceedings of the Biennial Meeting of the Philosophy of Science Association*, pp. 302–308.

Hunt, L. H. (2009), 'Literature as Fable, Fable as Argument', *Philosophy and Literature*, 33: 369–385.

Ivanova, M. (2017a), 'Aesthetic Values in Science', *Philosophy Compass*, 12, no. 10.

Ivanova, M. (2017b), 'Poincaré's Aesthetics of Science', *Synthese*, 194: 2581–2594.

John, E. (1998), 'Reading Fiction and Conceptual Knowledge: Philosophical Thought in Literary Context', *The Journal of Aesthetics and Art Criticism*, 56: 331–348.

Kapitsa, P. (1968), 'Reminiscences About Professor Ernest Rutherford' [1937], in A. Parry (ed.), *Peter Kapitsa on Life and Science*. Macmillan, pp. 75–99.

Levinson, J. (2010), 'Defending Hypothetical Intentionalism', *British Journal of Aesthetics*, 50: 139–150.

Matthews, R. J. (1977), 'Describing and Interpreting a Work of Art', *The Journal of Aesthetics and Art Criticism*, 36: 5–14.

McAllister, J. (1996), *Beauty and Revolution in Science*. Cornell University Press.

Meynell, L. (2018), 'Images and Imagination in Thought Experiments', in M. T. Stuart, Y. J. H. Fehige and J. R. Brown (eds.), *The Routledge Companion to Thought Experiments*. Routledge.

Newton, I. (1999 [1687]), *The Principia: Mathematical Principles of Natural Philosophy*, trans. I. B. Cohen and A. Whitman. University of California Press.

Norton, J. D. (2004), 'Why Thought Experiments Do Not Transcend Empiricism', in C. Hitchcock (ed.), *Contemporary Debates in the Philosophy of Science*. Blackwell, pp. 44–66.

Norton, J. D. (2018), 'The Worst Thought Experiment', in M. T. Stuart, Y. J. H. Fehige and J. R. Brown (eds.), *The Routledge Companion to Thought Experiments*. Routledge.

Nozick, R. (1974), *Anarchy, State and Utopia*. Blackwell.

O'Loughlin, I. and McCallum, K. (2019), 'The Aesthetics of Theory Selection and the Logics of Art', *Philosophy of Science*, 86: 325–343.

Palmerino, C. R. (2018), 'Discussing What Would Happen: The Role of Thought Experiments in Galileo's Dialogues', *Philosophy of Science*, 85: 906–918.

Parsons, G. G. and Rueger, A. (2000), 'The Epistemic Significance of Appreciating Experiments Aesthetically', *British Journal of Aesthetics*, 40: 407–423.

Schickore, J. (2018), 'Scientific Discovery', in E. N. Zalta (ed.), *The Stanford Encyclopedia of Philosophy*, Summer 2018 ed., Metaphysics Research Lab, Stanford University. Online at https://plato.stanford.edu/archives/sum2018/entries/scientific-discovery/.

Sibley, F. (1959), 'Aesthetic Concepts', *Philosophical Review*, 68: 421–450.

Stecker, R. (1994), 'Art Interpretation', *The Journal of Aesthetics and Art Criticism*, 52: 193–206.

Stecker, R. (2010), 'Do All Valuable Artworks Possess Aesthetic Value?', *Annales Philosophici*, 1.

Stuart, M. T. (2016), 'Taming Theory With Thought Experiments: Understanding and Scientific Progress', *Studies in History and Philosophy of Science Part A*, 58: 24–33.
Stuart, M. T. (2018), 'How Thought Experiments Increase Understanding', in M. T. Stuart, Y. J. H. Fehige and J. R. Brown (eds.), *The Routledge Companion to Thought Experiments*. Routledge.
Todd, C. S. (2008), 'Unmasking the Truth Beneath the Beauty: Why the Supposed Aesthetic Judgements Made in Science May Not Be Aesthetic at All', *International Studies in the Philosophy of Science*, 22: 61–79.
Vorms, M. (2011), 'Representing With Imaginary Models: Formats Matter', *Studies in History and Philosophy of Science Part A*, 42: 287–295.
Weatherson, B. (2010), 'Surveys and Thought Experiments', *Thoughts Arguments and Rants* (blog), August 24, 2010. Accessed March 27, 2019 online at http://tar.weatherson.org/2010/08/24/surveys-and-thought-experiments/.
Weinberg, J. M. (2008), 'Configuring the Cognitive Imagination', in K. Stock and K. Thomson-Jones (eds.), *New Waves in Aesthetics*. Palgrave Macmillan, pp. 203–223.

9 Epistemic Radicals and the Vice of Arrogance as a Counterfeit to the Virtue of Assured Epistemic Ambition

Matthew Kieran

1 Epistemic Radicals

Epistemic radicals play key roles in conceptual revolutions and technological innovation. At a much lower level, everyday epistemic transformations in disciplines, departments or projects often depend on people being epistemically radical. Even if this requires something like transformative or radical creativity (Boden 2004), the ability to be epistemically creative or radical is bound up with a cluster of features concerning epistemic character (Kieran 2019). This is to be expected especially if we take creativity in general to be tied up with aspects of an agent's character in particular concerning motivations, virtues and vices (Kieran 2014a, 2014b, 2018). Features of epistemic character such as epistemic ambition, unconventionality, resoluteness in the face of disagreement and resilience in the face of set-backs partly explain how radicals come to pursue seemingly unlikely possibilities, question disciplinary matrices, explore neglected conceptual spaces, entertain unorthodox assumptions and resolutely pursue inquiry in the face of epistemic indifference, ridicule, disdain and fundamental set-backs. Lynn Margulis, for example, met with "almost universal disbelief and scorn" (Ruse 2013) when arguing for the serial endosymbiotic theory of eukaryotic cell development. One rather emblematic evaluation concluded "your research is crap. Don't ever bother to apply again" (Bybee 2012: 157). Margulis met with initial indifference, followed by near unanimous disagreement, epistemic ridicule and significant intellectual ostracism. Yet Margulis' epistemic resoluteness, ambition and resilience in pursuing her project meant that initial scientific heresy eventually transformed the theory of evolution.

The problem is that the features characteristic of epistemic radicals are often associated with epistemic vice. According to conciliationism or the equal weight view (Christensen 2009; Frances and Matheson 2018), perhaps the dominant position in the disagreement literature, the steadfastness characteristic of an epistemic radical is viciously irrational. Frances (2010, 2014, 2016) even argues that most philosophers can only stick with their philosophical commitments as 'renegades' on pain of irrationality or some other epistemic vice. The tension between epistemic virtue and epistemic radicalism has not gone unnoticed. In the philosophical literature Paternotte and Ivanova (2016), for example, argue that while standard epistemic virtues are beneficial for conventional science, certain vices, such as egoism, conformism and dogmatism, are beneficial for the pre-convergence phase of scientific inquiry. Historical case studies also present epistemic vice as playing a causally explanatory role in the achievements of epistemic radicals. Michael White's *Acid Tongues and Tranquil Dreamers*, to take one example, details eight great rivalries in the advancement of science and technology as being fueled by intellectual arrogance, ego and competitiveness. To take another example, Sylvia Nasar, John Nash's biographer, states that at Princeton the "young mathematicians were all pretty cocky, but he [Nash] towered over them in arrogance and confidence and also in eccentricity" (Samuels 2002). According to Nasar, Nash took himself to be the measure of others, was frequently dismissive of anyone who disagreed with him and prided himself on being a free thinker who would work things out for himself (Nasar 2001: 67–69). He was also highly epistemically ambitious and possessed an "uncommon measure of self-confidence and self-importance. On one occasion, not long after his arrival at Princeton [as a student], he went to see Einstein and sketched some ideas he had for amending quantum theory" (Nasar 2001: 69–70). According to Nasar "no one was more obsessed with originality, more disdainful of authority, or more jealous of his independence . . . Even as a student, his indifference to others' skepticism, doubt and ridicule was awesome" (Nasar 2001: 12). Nash's intellectual arrogance—or at least something very close to it—is presented as helping to explain his epistemic achievements. There is even testimony from some epistemic radicals that arrogance is epistemically speaking a good thing. James Watson, for example, states that in scientific inquiry "you're not supposed to be arrogant, but if you're not arrogant, if you don't believe you know how to do something better than someone else, you're probably not doing anything" (Watson 1991).

Empirical work seems consonant with the thought that arrogance in particular makes for epistemic radicals. Feist's (1993) interview study of 100 leading scientists, blindly and independently evaluated, found "positive relations between ratings of hostility and arrogance and scientific eminence. The most eminent were deemed the most hostile and arrogant." Feist's subsequent metanalysis of creative personalities (1998) identified the aforementioned traits to be indicative of high creativity and, as he later summarized matters, "the traits of arrogance, hostility and conscientiousness (or relative lack therof) are most noteworthy of highly creative scientists. The confidence found in scientists in general seems to go one step further in the most creative scientists" (2006: 122). More recently, a study of 1,304 subjects using the HEXACO-60 found "people lower in Honesty-Humility had higher creativity scores consistent with past work on arrogance and pretentiousness among creative people" (Silvia, Kaufman, Reiter-Palmon and Wigert 2011: 687).

Now virtue can sometimes have bad consequences and vice good ones. However, this is supposed to be so only atypically and accidentally. Epistemic virtue is supposed to tend systematically toward epistemic good and vice toward epistemic failure and handicapping. Yet if the above is right, we have a significant tension between a) epistemic vice systematically tends toward being epistemically radical, b) epistemic virtue systematically impedes being epistemically radical and c) being epistemically radical is a highly valuable epistemic good. How should we resolve the tension?

In what follows this puzzle will be addressed by:

i) Showing how epistemic arrogance renders people susceptible to certain kinds of epistemic error and misdirection.
ii) Arguing that there is a nearby counterpart virtue of assured epistemic ambition that is insulated from those very susceptibilities.
iii) Showing how i) and ii) have significantly overlapping epistemic behavioural profiles such that arrogance is best conceived as a counterfeit virtue. This captures the complex relations arrogance has in relation to genuine virtue thereby explaining why people may often misattribute epistemic virtue to the arrogant and the epistemic vice of arrogance to the genuinely virtuous.
iv) Arguing that given i)—iii) epistemic radicals can be heroes or villains i.e. epistemically virtuous or vicious. Conceptualizing arrogance as an epistemic vice standing in a counterfeit relation to the true epistemic virtue of assured epistemic ambition shows how and why this is so.

2 The Strengths and Weaknesses of Epistemic Arrogance

Spinoza characterizes arrogant pride as "thinking more highly of oneself than is just, as a function of self-love" (Spinoza 2001 [1677]), Pt 3, Definitions of the Emotions XXVIII). Contemporary accounts vary in how exactly this core thought is cashed out:

a) Roberts and Wood (2007: 243–250) consider arrogance most fundamentally to involve a high sense of superiority motivating the disposition to infer false or illicit entitlement.
b) Whitcomb, Battaly, Baehr and Howard-Snyder (2017: 530–531) take arrogance to involve excessive self-attentiveness to strengths and a variety of 'over owning' dispositions including self-attributive tendencies to over-estimate, over-emphasize and over-attribute positive outcomes.

c) Tiberius and Walker (1998) argue that arrogance is constituted by a high self-opinion taken to legitimize the presumption of superiority qua human specimen which, in turn, generates tendencies to look down on others, dismissiveness and a failure to consider others' viewpoints.
d) Tanesini takes distinct kinds of arrogance to tend toward dismissiveness or the ignoring of others and an 'unwillingness to submit oneself to the norms governing ordinary conversation and rational debate' (Tanesini 2016: 85) including tendencies toward self-exemption from the requirement to justify claims (Tanesini 2018a).

On all major accounts, arrogance is bound up with characteristics that enable someone to be an epistemic radical. An excessive attention to epistemic strengths and associated dispositions tends to generate greater ambition and reinforce the drive to realize such. Where someone is disposed to infer illicit entitlement it is much easier to make assumptions or work from commitments that go beyond whatever is justified by the available evidence. Where someone tends to be epistemically dismissive, ignore the work of others and feels less bound by norms governing rational debate, then it is much easier to be resolute in the face of disagreement. Note that being epistemically arrogant is consistent with possessing certain epistemic virtues even to a high degree (e.g. curiosity). Hence epistemic radicals can be epistemically virtuous in certain respects even while being vicious in another (i.e. epistemically arrogant). Nonetheless, arrogance by its nature systematically tends toward epistemic failure. What is the maxim or rule under which the epistemically arrogant act? '*My* superior epistemic position ensures success or insight'. Notice that this connects with Tiberius and Walker's (1998: 382) characterization of arrogance as the presumption that one is 'better qua human specimen' while recognizing that the particular presumption need not be involved.

The Neo-Darwinians who ridiculed Margulis, for example, may just have been complacent about the (falsely) presumed superiority of their epistemic vantage point. As Margulis characterizes matters the fact that many of them came from zoology led to an overly narrow focus on and over-generalization from the limited case of animals. This looks like a case of inferring illicit entitlement ala Roberts and Wood (2007) or excessive attention to the strengths of a position ala Whitcomb et al. (2017) which led the Neo-Darwinians to make overly general claims from too narrow an evidential base. In effect the upshot was to codify ignorance given that their approach misses "four out of the five kingdoms of life. Animals are only one of these kingdoms. They miss bacteria, protoctista, funghi, and plants. They take a small and interesting chapter in the book of evolution and extrapolate it into the entire encyclopedia of life. Skewed and limited in their perspective, they are not wrong so much as grossly uninformed" (Margulis 1996: 130). The fact that the Neo-Darwinians were so dismissive of Margulis's arguments compounded by the vitriolic ridicule involved is also consonant with Tanesini's (2016, 2018a) characterization of arrogance as trampling roughshod over the standard norms of discourse in ways bound up with superiority, dismissiveness of others and exemption from epistemic justification.

Arrogance's self-presumption generates reckless ambition manifested in carelessness over methods, short cuts in approach and epistemic licence. This guiding modus operandi in turn explains how even early problematic signals are ignored, set backs rationalized away and epistemic norms disregarded. As a result, epistemic projects will often crash and burn. Lysenko's overreaching epistemic ambition was bound up with arrogation from the norms of decent science. Infamously Lysenko was culpably reckless in method, viciously dismissive of criticism and over generalized from experiments far too quickly (Joravsky 1970; Graham 2016). In the more recent case of Diederik Stapel we have an eminent psychologist who faked a lot of his data seemingly to 'prove' what he 'knew' anyway (Bhattacharjee 2013; Levelt, Noort and Drenth 2012). Retractions of Stapel's work in peer reviewed journals at the last count as identified by Retraction Watch came to 58 articles (Retraction Watch 2015). Less dramatically, arrogance often tends toward epistemic rigidity, speculation taken as certainty, data cherry picking, over generalization, over claiming and obstinacy in the face of criticism. Where inquirers arrogantly presume their conceptualization or approach must be right, they often fail to see otherwise obvious phenomena, problems or alternative explanations that fail to mesh with their projected schemas. This leads to a kind of insensitive deliberation and feedback failure, which in turn only serves to reinforce close minded stubbornness.

3 The Virtue of Assured Epistemic Ambition

There is, however, a counterpart virtue to the vice of arrogance: the virtue of assured epistemic ambition. This can be characterized as follows:

> S possesses assured epistemic ambition to the degree that S is admirable in i) having high epistemic ambitions ii) which are internalized as being immensely valuable or valuable for their own sake iii) and is committed to pursuing them appropriately with iv) an epistemically permissible high degree of self-trust in presuming she has a good enough chance at realizing them.

To be virtuous the epistemically ambitious must strive for great epistemic achievement either for its own sake or because it is justifiably taken to have significant value. The motivation alone is insufficient for the virtue since it must be rationally permissible for the person to think that she is capable of succeeding. This entails that the agent must have rationally permissible high self-belief that she is or could become capable of realizing her high epistemic ambitions in some form. The characterization picks out people with high epistemic ambitions who are committed to realizing them. They back themselves to give the relevant project or enquiry a good go. It is epistemically permissible to do so given what they are justified in believing, their strengths, what may justifiably be made of the present state of evidence and the possible prospects for the epistemic project. Notice that such agents need not be aware of or think of themselves in self-approving ways. While the characterization is consistent with people thinking highly of themselves for being like this, self-approbation is not part of what it is to possess assured epistemic ambition. We admire people with great epistemic ambitions. What is crucial here is that such agents take themselves to be epistemically ambitious for epistemic goods that are either valuable for their own sake or for further goods that are immensely valuable. This goes a long way to explaining why such people tend to remain committed to their inquiries even where doing so results in indifference, ridicule and ostracism.

Exactly how the ambitions are construed as valuable may be psychologically configured in a variety of ways. People can take their high epistemic ambitions to be immensely valuable for their own sake. In other words, what matters is working out what exactly is significantly puzzling in the relevant domain, how to explore the terrain, experiment, conceptualize matters and try out potentially valuable solutions. The discovery of polonium then radium by the Curies, for example, seems to be the story of two scientific idealists driven on by epistemic curiosity, fascination and the epistemic drive for knowledge for its own sake (Curie 2012 [1923]; Pasachoff 1996). Having epistemic ambitions taken as valuable for their own sake is, note, compatible with having further reasons (which may lead to symbiotic reinforcement). People can be interested in some domain or set of questions for the sake of knowledge while possessing additional reasons for being drawn into an area of inquiry in the first place. Jane Goodall both loved animals from an early age and was fascinated by them. By the age of ten she was dreaming of living with animals in Africa (Goodall 2001; Greene 2005). The symbiotic interaction of Goodall's early love of animals and epistemic curiosity helps to explain her radical departures in observation and methods such as her naming and descriptions of individual animals (Goodall 2001, 2002). Her epistemic motivations came to dovetail for a large part of her life with the partly non-epistemic conservationist or environmentalist motivations that came to figure so largely in her life. In some cases, knowledge for its own sake may not figure at all given that what is taken to matter is epistemic significance for the sake of some further non-epistemic goal. Donald Hopkins played a leading role in wiping out smallpox in Central and West Africa and then went on to play an even greater role in eradicating the now near extinct guinea worm disease. As a Morehouse College chemistry undergraduate he went to the Institute of European Studies and then travelled more widely. In Egypt he was struck by just how severe and widespread eye infections were (Oakes 2000: 347). Hopkins decided "then and there that I wanted to work on tropical diseases" to alleviate human suffering (PBS). He returned home, worked hard at being transferred to the University of Chicago to study medicine, became the only black person to graduate in his cohort (Yeoman 2017), and then devoted his life to eradicating infectious diseases.

The virtue of assured epistemic ambition may involve—though need not—desiring epistemic achievement for its own sake. Nonetheless, people can possess the relevant dispositions constituting the epistemic virtue while being motivated to realize high epistemic ambitions for some further, ultimate end such as relieving suffering, bringing about a more just society or enabling space travel. Even so, there are minimal constraints on how this can be so in order to constitute the virtue:

a) The agent must possess the sincere, epistemically permissible belief that the line being explored is or might be a good way of realizing the ultimate ends aimed at.
b) The ultimate ends being aimed at must have significant, immense value.
c) The inquiry must be pursued in a particular non-wholly instrumentalized way. To be more specific the agent's epistemic conduct must respect and honour proper epistemic constraints, duties, permissions and values.

Consider a basic contrast. A person may pursue her scientific inquiry for the sake of making people's lives better in some way. She sincerely, justifiably believes that there are decent grounds for pursuing the line of inquiry. Yet in conducting her inquiry, she fails to do justice to the standards and values of decent epistemic investigation. This might be manifest in a whole host of ways such as being culpably careless in not running certain tests, in failing to ensure proper experimental conditions, cherry picking data, dismissing negative results, filing away inconclusive data, or even in extremis making things up. Egotistical self-promotion and careerism are further ends often taken to explain such failings. Yet this can be the case even where the fundamental driving motivation is beneficent. Even if non-epistemic admirable, beneficent ends had been driving Lysenko, his failure to respect epistemic constraints, norms and the procedures of good science would have been epistemically vicious. People can commit blameworthy, vicious epistemic failings in part because they are psychologically overcommitted in the wrong kind of way to what they wish to be the case. The epistemically virtuous, by contrast, seek to do justice to strictly epistemic constraints and abide by epistemic norms no matter how much it matters that the value of what they are doing depends on realizing the non-epistemic worthy end goal. The recognition of this point only requires cognizance of the fact that there is a hierarchy of motivational structure to our actions. An epistemic action or disposition can be virtuous, in respecting the internal nature of epistemic duties and goods, while nonetheless being ultimately for the sake of some further final goal or end. This way of putting matters is intended to be as neutral as possible between various distinct ways of cashing out the notions of doing something for its own sake and for the sake of something else (for different kinds of account see, for example, MacIntyre 2007: 181–203; Rabinowicz and Ronnow-Rasmussen 2015: 30–34; Williams 2002) though it does suggest that epistemic virtue, contra responsibilism (Zagzebski 1996: 165–197), does not have to involve the ultimate motivation of pursuing knowledge for its own sake (Baehr 2018; Kieran 2019).

The virtue requires not just possessing epistemic ambitions, but the right kind of commitment to them. The virtuous are driven to seek out, explore and take on the means of realizing their ambitions appropriately: they work at skilling up, acquiring expertise, cast around for interesting problems, look to address difficult challenges, avoid giving themselves easy passes and persevere in the face of set-backs. Crucially it is not enough for people just to back an epistemic project. The virtuous have to trust in the project and themselves. Hence thoughts like 'why not?', 'this has a decent shot' or 'this looks like a promising way to go' in pursuing difficult inquiry have to be epistemically permissible rather than unjustifiably misplaced or deluded.

People who possess assured epistemic ambition often set themselves at what sometimes seems to their peers to be unpromising or less than fully justified lines of inquiry. This is part of what makes the ambitiously assured epistemic risk takers. Yet the ambitiously assured are far from deluded since they are prudent risk takers relative to the nature of their epistemic ambitions. We would not admire epistemically ambitious agents seeking to taxonomize the kinds of leprechauns to be found at the ends of rainbows. Minimally, it must be epistemically permissible for the ambitiously assured to think that pursuing the line of inquiry they are committed to has a shot at realizing the relevant epistemic ambitions.

We can put the point rather differently. People who possess the virtue of assured epistemic ambition tend to take risks, explore neglected avenues and pursue new approaches where doing so justifiably looks like a genuine, worthwhile epistemic possibility. Even if nearly everyone else is dismissive based on the evidence they attend to or their disciplinary conceptual schemes, there must be something from an epistemic point of view which renders the approach promising even if the hope seems a comparatively remote one (at least in the initial stages at any rate). Minimally, then, those with assured epistemic ambition pursue enquiry on the basis of the warranted hope that something is a live option rather than, say, on the basis of deluded optimism. Genuine hope is rational and so, for example, entertains the possibility of failure in ways that optimism in virtue of its nonrational or irrational nature does not (Eagleton 2015). To the degree that someone possesses assured epistemic ambition she has the ability to take on difficult, challenging inquiries or new approaches in pursuit of the valuable high epistemic goals that she is committed to and—justifiably—trusts herself in doing so. It does not thereby follow in so self-trusting that she takes herself to be utterly reliable or bound to succeed (Hawley 2018). The default question that those with the virtue of assured epistemic ambition ask with respect to great epistemic ambitions is 'is this is a good way to go?' as contrasted with the presumption of the arrogant who just assume 'I am bound to succeed'. Hence assured epistemic ambition is in principle open to and watchful over the possibility of failure as contrasted with the complacency of arrogance.

The characterization given helps to explain a virtuous cycle of epistemic self-development. In the early stages someone with such a disposition will tend to approach difficult but reachable challenges and, in so doing, skill up and develop her capacities more quickly than the less ambitious. This, in turn, will make more difficult challenges more approachable, and she will come to gain greater justification for thinking that she can overcome difficult challenges and realize high ambitions. Given that she is open to the possibility of failing she will learn to deal with and watch out for failure as a matter of prudence without thereby dampening her great ambitions. The net result over time is an ever-increasing development of ability that puts her in a better position to realize great ambitions. The cycle gives the possessor increasing assurance and trust in her epistemic commitments, abilities and judgement when faced with indifference and disagreement. Hence the possessor tends toward fortitude and courage in epistemic inquiries, such as that displayed by the Curies and Margulis, in the face of challenges, ridicule and hostility from others. This helps to explain close links between assured epistemic ambition and other important epistemic virtues.

Contemporary philosophical literature often takes genuine or proper pride to be the contrasting virtue to both arrogance and servility (see, for example, Whitcomb et al. 2017: 530; Tanesini 2018b). However, assured epistemic ambition is distinct from proper pride (though consistent with it). Why? People without large ambitions often can and do possess genuine, merited pride. More importantly, pride necessarily involves self-approbation in a way that epistemic assurance does not. You may trust yourself to write that really hard book, come up with a creative experiment or develop some theory without involving any degree of self-approbation. Assured epistemic ambition focuses on doing the project or the inquiry (i.e. is outward focused) whereas pride necessarily involves self-consciously approving attitudes (i.e. is approvingly self-directed). The proud will tend to possess high-self confidence and think 'I did that' or 'I can do this' in self-approving terms. The ambitiously assured will tend to think 'this idea is worth pursuing' or 'that line of inquiry looks promising'. The proud self-consciously approve of and have confidence in themselves, the ambitiously epistemically assured have confidence in the idea or inquiry and trust themselves to pursue it.

As characterized assured epistemic ambition has features that place people in a good position to be epistemic radicals without being susceptible to some of arrogance's error or misguidance tendencies. In particular the self-assured will be more:

a) open to the possibility of failure and so less inclined to be complacent about the nature of epistemic challenges e.g. more open to early set backs or failures as signals for things to work on.
b) open to disagreement as something worth taking seriously rather than being a function of others inferiority.
c) co-operative for the ambitiously assured the idea is the thing whereas for the arrogant the presumption of superiority is the thing. Hence people with assured epistemic ambition tend to be more open than the arrogant in recognizing that good ideas or help can come from others (and can often do so irrespective of status).

Before moving on to consider the relationship between assured epistemic ambition and arrogance, it is worth saying something about a contrast with epistemic humility. Humility is taken constitutively to involve dispositions to attend to, consider and acknowledge failings, limitations and weaknesses (Whitcomb et al. 2017). Furthermore, in classical terms at least, it is taken to be contrary to humility to aim at great things by trusting in our own powers (Aquinas 1947: ST II-II, 161, A. 1). This helps to explain why Hume denounced humility as a 'monkish' virtue and it is notable that nothing close to humility figures as a virtue for Aristotle. On the above characterization assured epistemic ambition is partly constituted by aiming at great things in ways involving epistemic self-trust. It does not necessarily follow, however, that assured epistemic ambition is incompatible with humility. As Aquinas characterizes matters, humility is "concerned to temper and restrain the mind, lest it tend to high things immoderately" (Aquinas 1947: ST II-II, 161, A. 1). Humility thus serves to correct and rein in inappropriate or immoderate epistemic ambition. Much will turn on what 'immoderate' here amounts to. Nonetheless, humility might thus be in principle consistent with, though distinct from, assured epistemic ambition given the latter constitutively involves the appropriate pursuit of high epistemic ambitions and trust in your epistemic potential, powers and strengths to do so. Hence the virtuous pursuit of ambitious enquiry can tend toward ambitious risk taking without involving the recklessness born of arrogance. Arrogance is closed off from and complacent about possibilities of error, misguidance and owning epistemic limitations. By contrast, assured epistemic ambition is open to such possibilities though openness to such may be distinct from fully recognizing or owning epistemic weaknesses (which might constitute the epistemic virtue of humility).

4 Arrogance as a Counterfeit Virtue

Nonetheless, in many instances people who possess the virtue of assured epistemic ambition and those who possess the vice of arrogance may end up with apparently similar, overlapping behavioural profiles. Take, for example, judgements with respect to epistemic peerhood. As characterized above, people with assured epistemic ambition not only set their epistemic sights high, but come to acquire a comparatively high degree of technical skills, knowledge and understanding. It follows that they may often regard only a comparatively select few people as epistemic peers. This follows from the fact that they have good justification for holding that they have a greater degree of knowledge or understanding than most others in the relevant domains. As observed from the third person point of view, those with assured epistemic ambition can thus seem remarkably close to the behavioural profile of the epistemically arrogant. After all, the epistemically arrogant tend not to recognize many others as their epistemic peers and so fail to take them seriously. Furthermore, where possessors of the virtue and the vice judge that a peer's testimony or disagreement is worth taking seriously, they'll all be strongly motivated to investigate for themselves rather than just accept the testimony or reasoning of others (even if only as something to think about and overcome in the project or inquiry). In other words, the epistemic benefits of the virtue as contrasted with the epistemic susceptibilities of the vice might not show up much in many instances of judgements of peerhood and epistemic interactions with others (though they will do so in other contexts). This is not to say that there are no differences. The dispositional patterning in principle will tend toward certain key differences. The arrogant, for example, will have a much stronger tendency to take lack of interest or disagreement as such as evidence of epistemic inferiority, whereas the self-assured tend to be more epistemically open to learning from others (whatever their epistemic status) and accepting of differences in epistemic interests. The epistemically assured, for example, will judge someone merely to have different epistemic interests in certain instances where the arrogant would dismiss someone as epistemically inferior. Nonetheless, assured epistemic ambition can often be mistaken for arrogance given that both have tendencies towards features such as:

i) independence of mind
ii) interest in and commitment to ambitious projects and inquiries
iii) resoluteness in the face of disagreement

Such strengths help to explain why both those with assured epistemic ambition and the arrogant often make for epistemic radicals. Crucially, however, those possessed of assured epistemic ambition have strengths the arrogant lack and lack failings that the arrogant are susceptible to. Consider, by way of example, epistemic arrogance's failings such as inattentiveness to possible failings or challenges, insufficient motivation to address such, rigid close mindedness and a presumption of entitlement to success (which in extremis can lead to abrogating the norms of epistemically good inquiry).

If the above is right, then arrogance might best be thought of as a vice that is the counterfeit virtue for the true virtue of assured epistemic ambition. What is the notion of a counterfeit virtue? Aquinas makes use of the notion in reminding us that a proper virtue is orientated toward a good that is an end though actions may have the semblance of virtue in being orientated only toward or possessing the semblance of the good in which case "it is not a true virtue that is ordered to such a good, but a counterfeit virtue" (Aquinas 1947: ST II-II, q. 23, a. 7). In more general terms, a counterfeit is something that can pass for something else while lacking some key feature or relation required to constitute the genuine article. Now, counterfeit money or goods, for example, often can and do pass extremely well for the real thing. Good counterfeits can be used to serve many of the same functions as the genuine article ranging from utility value to social signaling (in many though not all circumstances).

What makes arrogance a good *counterfeit* for assured epistemic ambition? There will be a large degree of overlap between the epistemic behavioural profiles of the arrogant and those with assured epistemic ambition. The profiles will not be identical, for the reasons given above, but the epistemically arrogant often have high epistemic ambitions, are prepared to take many similar kinds of risks, are independent minded and forge ahead in the face of indifference, disagreement and ridicule. Transformative epistemic creativity, and the creative development required to become an epistemic radical typically depends upon a wide range of traits and behaviours such as initiative, risk taking, opportunity seeking, persistence, boldness, assertiveness, daring, resilience in the face of failure, single mindedness, self reliance, self belief and the abilities to deal with a high degree of uncertainty and any associated anxiety over extended periods of time. In many circumstances both assured epistemic ambition and arrogance can underwrite or give rise to such behaviours.

It is worth pointing out several practical implications here. First, if we consider the empirical work cited above (Feist 1993, 1998, 2006; Silvia et al. 2011), there is reason to think that the operationalization of arrogance (or lack of humility) is too broad. It could be that the arrogant and those with assured epistemic ambition are being falsely conflated together as arrogant. Second, notice that where people are overly deferential, fearful, self-doubting, people pleasing, weak in the face of disagreement or epistemically insecure, a useful practical heuristic for them might be to aim to be 'more arrogant'. This is not because in aiming to be more arrogant such people will likely hit the behaviour profile of the arrogant. Rather, given where such characters start from, in aiming to be a bit more arrogant, as they would see it, they will likely come closer to the behavioural profile of the more self-assured (and in so doing may start to build up greater confidence and ability). Nonetheless this can be a tricky thing since as we have seen there is a fine line between assured epistemic ambition and arrogance.

It is important to realize that the counterfeit relation works both ways. This has certain practical consequences such as the perpetration of epistemic injustice where people are wronged in their capacities as knowers or epistemic agents (Fricker 2007). Recognizing that arrogance is the counterfeit virtue to the genuine virtue of assured epistemic ambition helps to explain why the epistemically vicious may easily be mistaken for the epistemically virtuous and vice versa. This possibility is likely compounded by interaction with certain moderating factors such as stereotype effects. People from disadvantaged groups for example, may often be much more easily construed as arrogant when they manifest the epistemic profile of assured epistemic ambition precisely because the profile is in tension with the stereotypically assumed profile qua member of that group. Women who are ambitiously assured in domains such as mathematics, physics or philosophy may be condemned as arrogant because the profile of the virtue is in tension with gendered concepts concerning how women 'are' or are 'supposed to be'. Hence members of the relevant group may be punished when they display the epistemic virtue (and one way of punishing people is to condemn them as vicious). Conversely, when members from an advantaged group are being arrogant this might be more easily mistakenly for the virtue of assured epistemic ambition where the profile is consonant with relevant stereotypical assumptions. Recognizing that arrogance is a counterfeit virtue for assured epistemic ambition may thus help to deepen our understanding of how epistemic injustice works in certain contexts. After all, cultivating assured epistemic ambition may thus often be much more difficult—and so a much more challenging achievement—for members of disadvantaged groups. Moreover, the obvious pursuit of the virtue in certain domains may be an apparently imprudent strategy for members of disadvantaged groups unless they are prepared to cope with or confront a host of difficulties.

Nonetheless, it might be objected that in such scenarios being arrogant will lead to greater epistemic benefit for some small number of individuals (though this may go along with greater epistemic disbenefit for a far greater number of individuals). Furthermore, the overall net gain to radical, innovative science may be epistemically beneficial. Hence, at least to the extent that we prize epistemic radicals and innovative science, we would want some number of people to be arrogant. Hence arrogance cannot be a true epistemic vice and assured epistemic ambition cannot be a true epistemic virtue.

The objection misses the mark because it fails to take into account the normative aspect of assured epistemic ambition. This is about how to be a good, admirable epistemic agent. It could be in certain problematic environments that the epistemic benefits of arrogance are more obvious and even exceed those of the virtue under particular circumstances. What this would then show is that something is wrong with the way the epistemic environment is. Consider a putative sub-culture of inquiry driven by assumptions and norms tied to competitive, individualist 'star' systems tracking self-confident performance. Such environments might tend to cultivate or compound arrogance taken as ambitious assurance. This might have a form that fits the following basic schematic:

i) Successful 'stars' give advice, tips and model success along the lines of 'do as I did'. They do so without recognizing that much of what they did was high risk that they got lucky and 'their' work was highly dependent on others. Hence the individuals in question have strong tendencies to overattribute responsibility for success to themselves rather than to the work of others as well as good luck.
ii) Inquirers act within structural or institutional systems that strongly promote high risk taking. By way of example, people may be promoted, honoured and awarded significant research grants only when they have big successes. Hence only the very high risk seeking are disproportionately promoted or honoured, even if it is also the case that a very large number of high risk takers fail (or have to leave in disgrace).

Even without any further complicating factors if i) is combined with ii) the upshot might be a simple psychological recipe for inculcating and rewarding arrogance (at least at the 'higher end' of structural hierarchies and organizational ranks). What this would show is not that arrogance is not a true epistemic vice in such a scenario but, rather, that there is something problematic about the epistemic environment.

A rather different objection might articulate the thought that at least much of the time we surely do not want all inquirers to be radical. A lot of inquiry is developed through working out, elaborating and refining ideas in rather conventional ways. If assured epistemic ambition is something that we want people to possess only at particular epistemic junctures, or something we want only some small number of inquirers to possess, then it cannot be a virtue. Work in the social epistemology literature might seem consonant with just this thought. It has been argued not just that diversity in research approach in a scientific population is more epistemically efficient (Zollman 2010), but that the most successful epistemic communities (i.e. the most efficient groups capable of realizing their epistemic goals more quickly), are constituted by two different groups of scientists: mavericks and followers (Weisberg and Muldoon 2009). In such communities only a small number of groups of mavericks are required to generate new ideas that are then developed by the followers. It has, further, been argued that mavericks are arrogant, egotistic, self-centered and focus on their own ideas, while followers are more modest or humble in exploring and refining the ideas of the mavericks (Paternotte and Ivanova 2016).

The first thing to say here is that just because something is an epistemic virtue, it does not follow that it must be manifested. What it is to possess a virtue involves being disposed to manifest patternings of thought, affect and action appropriately in the relevant virtue eliciting conditions. It may be that many people much of the time are not in the appropriate epistemic circumstances to pursue ambitiously unconventional epistemic inquiries. Consider, by analogy, non-epistemic courage. If anything is a virtue, then courage is. Yet it may be that in good conditions for much of the time many people are not required to be courageous. The truly virtuous will be courageous only when courage is called for. The same can be said for the epistemic virtue of assured epistemic ambition. It may be that great epistemic ambition is not always or perhaps not even often called for given the relevant situation.

The second thing to say is that if the argument presented in this chapter holds, then it is a mistake to hold that mavericks must tend toward vices such as arrogance, egotism and self-centredness. Rather, as has been argued, the true virtue of assured epistemic ambition affords the relevant epistemic goods more consistently, reliably and without the weaknesses of the counterpart vices mentioned.

The third thing to say is that the virtue theorist can opt for either a more global response to the worry—the one I tend toward myself—or a more situationally specific response.

The global virtue theorist will hold that assured epistemic ambition is an epistemic virtue that is partly constitutive of what it is to be a good inquirer. Everyone should possess the virtue. As noted, it does not necessarily follow that this should be manifested all or much of the time. Nonetheless, any good inquirer at some point in pursuit of their epistemic interests will likely need to call on the virtue—at least if their epistemic endeavours are to make any headway. One way to make this view appealing is to consider just what it is to develop as an epistemic agent. There is a tendency to think of epistemic development as a kind of linear progression in the development of epistemic skills and understanding. But this is far from being true. If anything, we must unlearn certain habits and learn how to question or radically reconceptualize key assumptions. This might not be the most radical thing to do in the fullest degree, but it does require at least a minimal degree of assured epistemic ambition. Hence the virtue is required to some degree both in developing as an epistemic agent and at certain stages in pretty much any project or inquiry. The virtue will be exercised as, when and where it is required.

The more situationally specific response, by contrast, will acknowledge that we do not want or need most inquirers to possess ambitious assurance. We only need some individuals or teams to have such a character. Note that a team can possess the overall composite character of being ambitiously assured without necessarily each or perhaps any individual being so. If it is a good idea for a particular group to be striving for ambitious assurance, it may be a fine art in balancing the comparative virtues and relational character of the composite individuals so that the team as a whole is ambitiously assured. In principle there might be many possible configurations that could give rise to a group possessing the virtue as a collective. More crucially, the more situationally specific response will hold that we would not want every or even most inquirers to have this particular virtue. It is one good way to be amongst others, though it is the best way to be if you—either as an individual or a collective—want to be epistemically radical. It is just that not all of us do or should want to be epistemic radicals. In principle we might need individual people and epistemic groupings who are epistemic moderates just as much as we need epistemic radicals.

5 Epistemic Radicals: Heroes and Villains

Thinking of assured epistemic ambition as the virtue to which arrogance pays the compliment of being a counterfeit gives us a more nuanced, faithful account of epistemic radicals. Epistemic radicals may be epistemically virtuous or vicious. They *can* be virtuous in possessing the virtue of assured epistemic ambition, which puts them in a good place to realize high end, difficult, transformative epistemic goods (ones those lacking the virtue are less well placed to realize). Nonetheless, epistemic radicals as a matter of fact can be and sometimes are vicious in so far as they are epistemically arrogant. Arrogance can play a role in helping to explain causally how some epistemic radicals come to achieve epistemic goods while nonetheless not being the only or the best way to realize them. Even in such cases, epistemic arrogance, whether of an individual, group or epistemic culture more generally, always warrants condemnation. Moreover, epistemic arrogance always has certain error or misguidance susceptibilities, which the true virtue of assured epistemic ambition does not. Epistemic arrogance is at best a counterfeit to the true virtue of assured epistemic ambition.

Note

Versions of this chapter were presented at the University of Leeds at the Aesthetics and Science conference, July 2017, an earlier History and Philosophy of Science work in progress seminar, and the Creativity, Imagination and Rationality workshop at the University of Bristol in December 2016. Many thanks to all those present for helpful questions, comments and suggestions. In addition, many thanks for further helpful discussion, comments and suggestions to Victor Durà-Vilà, Catherine Elgin, Catarina Dutilh Novaes, Simon Hewitt, Robert Simpson and Robbie Williams.

Bibliography

Aquinas, T. (1947 [1265–74]), *Summae Theologica*, trans. Fathers of the English Dominican Province. New York: Benziger Bros.

Baehr, J. (2018), 'Intellectual Creativity', in B. Gaut and M. Kieran (eds.), *Creativity and Philosophy*. Abingdon: Routledge.

Bhattacharjee, Y. (2013), 'The Mind of a Con Man', *New York Times, Magazine*, April 26. Online at www.nytimes.com/2013/04/28/magazine/diederik-stapels-audacious-academic-fraud.html?pagewanted=all&mcubz=0

Boden, M. A. (2004), *The Creative Mind: Myths and Mechanisms*, 2nd ed. Routledge.

Bybee, J. (2012), 'No Subject Too Sacred', in D. Sagan (ed.), *Lynn Margulis: The Life and Legacy of a Scientific Rebel*. Vermont: Chelsea Green Publishing.

Christensen, D. (2009), 'Disagreement as Evidence: The Epistemology of Controversy', *Philosophy Compass*, 4: 756–767.

Curie, M. (2012 [1923]), *Pierre Curie With Autobiographical Notes*, trans. C. and V. Kellogg. Dover.

Eagleton, T. (2015), *Hope Without Optimism*. Yale University Press.

Feist, G. J. (1993), 'A Structural Model of Scientific Eminence', *Psychological Science*, 4: 366–371.

Feist, G. J. (1998), 'A Meta-Analysis of Personality and Scientific and Artistic Creativity', *Personality and Social Psychology Review*, 2: 290–309.

Feist, G. J. (2006), *The Psychology of Science and the Origins of the Scientific Mind*. Yale University Press.

Frances, B. (2010), 'The Reflective Epistemic Renegade', *Philosophy and Phenomenological Research*, 81: 419–463.

Frances, B. (2014), *Disagreement*. Oxford: Polity Press.

Frances, B. (2016), 'Worrisome Skepticism About Philosophy', *Episteme*, 13: 289–303.

Frances, B. and Matheson, J. (2018), 'Disagreement', in E. N. Zalta (ed.), *The Stanford Encyclopedia of Philosophy*, Spring 2018 ed. Online at https://plato.stanford.edu/archives/spr2018/entries/disagreement/.

Fricker, M. (2007), *Epistemic Injustice*. Oxford University Press.

Goodall, J. (2001), *Africa in My Blood: An Autobiography in Letters, the Early Years*, ed. D. Peterson. New York: First Mariner Books.

Goodall, J. (2002), *Beyond Innocence: An Autobiography in Letters, the Later Years*, ed. D. Peterson. New York: First Mariner Books.

Graham, L. (2016), *Lysenko's Ghost: Epigenetics and Russia*. Harvard University Press.

Greene, M. (2005), *Jane Goodall: A Biography*. Westport, CT: Greenwood.

Hawley, K. (2018), 'Creativity and Knowledge', in B. Gaut and M. Kieran (eds.), *Creativity and Philosophy*. Abingdon: Routledge.
Joravsky, D. (1970), *The Lysenko Affair*. Cambridge, MA: Harvard University Press.
Kieran, M. (2014a), 'Creativity as a Virtue of Character', in E. Paul and S. Kaufman (eds.), *The Philosophy of Creativity: New Essays*. Oxford: Oxford University Press.
Kieran, M. (2014b), 'Creativity, Virtue and the Challenges From Natural Talent, Ill-Being and Immorality', *Royal Institute of Philosophy Supplement*, 75: 203–230.
Kieran, M. (2018), 'Creativity, Vanity and Narcissism', in B. Gaut and M. Kieran (eds.), *Creativity and Philosophy*. Abingdon: Routledge.
Kieran, M. (2019), 'Creativity as an Epistemic Virtue', in H. Battaly (ed.), *Handbook of Virtue Epistemology*. Abingdon: Routledge.
Levelt, Noort and Drenth Committees. (2012), *Flawed Science: The Fraudulent Research Practices of Social Psychologist Diederik Stapel*. Tilburg: Commissioned by the Tilburg University, University of Amsterdam and the University of Groningen.
MacIntyre, A. (2007), *After Virtue*, 3rd ed. London: Duckworth.
Margulis, L. (1996), 'Gaia is a Tough Bitch', in J. Borckman (ed.), *The Third Culture: Beyond the Scientific Revolution*. New York: Touchstone.
Nasar, S. (2001), *A Beautiful Mind*, 2nd ed. New York: Simon and Schuster.
Oakes, E. H. (2000), *Encyclopedia of World Scientists*. New York: Facts on File Inc.
Pasachoff, N. (1996), *Marie Curie and the Science of Radioactivity*. Oxford: Oxford University Press.
Paternotte, C. and Ivanova, M. (2016), 'Virtues and Vices in Scientific Practice', *Synthese*, 194: 1787–1807.
PBS. (1998), 'Donald Hopkins', *A Science Odyssey: People and Discoveries*. Online at www.pbs.org/wgbh/aso/databank/entries/bmhopk.html
Rabinowicz, W. and Ronnow-Rasmussen, T. (2015), 'Value Taxonomy', in T. Brosch and D. Sander (eds.), *Handbook of Value: Perspectives From Economics, Neuroscience, Philosophy, Psychology and Sociology*. Oxford: Oxford University Press.
Retraction Watch. (2015), *Diederik Stapel Now Has 58 Retractions*. Online at https://retractionwatch.com/2015/12/08/diederik-stapel-now-has-58-retractions/
Roberts, R. C. and Wood, W. J. (2007), *Intellectual Virtues*. Oxford: Clarendon Press.
Ruse, M. (2013), 'Review of Lynn Margulis: The Life and Legacy of a Scientific Rebel by Sagan', *Quarterly Review of Biology*, 88: 233–234.
Sagan, D., ed. (2012), *Lynn Margulis: The Life and Legacy of a Scientific Rebel*. Vermont: Chelsea Green Publishing.
Samuels, M. (2002), *A Brilliant Madness*. [Documentary Film] American Experience PBS Series.
Silvia, P. J., Kaufman, J. C., Reiter-Palmon, R. and Wigert, B. (2011), 'Cantankerous Creativity: Honesty-Humility, Agreeableness and the HEXACO Structure of Creative Achievement', *Personality and Individual Differences*, 51: 687–689.
Spinoza, B. (2001 [1677]), *Ethics*. London: Wordsworth Editions.
Tanesini, A. (2016), '"Calm Down Dear": Intellectual Arrogance, Silencing and Ignorance', *Aristotelian Society Supplementary Volume*, 90: 71–92.

Tanesini, A. (2018a), 'Arrogance, Anger and Debate', *Symposium: Theoretical and Applied Inquiries in Philosophy and Social Sciences*, 5: 213–227.

Tanesini, A. (2018b), 'Intellectual Servility and Timidity', *Journal of Philosophical Research*, 43: 21–41.

Tiberius, V. and Walker, J. D. (1998), 'Arrogance', *American Philosophical Quarterly*, 35: 379–390.

Watson, J. (1991), *The Academy of Achievement Interview*. Online at www.achievement.org/achiever/james-d-watson/#interview.

Weisberg, M. and Muldoon, R. (2009), 'Epistemic Landscapes and the Division of Cognitive Labor', *Philosophy of Science*, 76: 225–252.

Whitcomb, D., Battaly, H., Baehr, J. and Howard-Snyder, D. (2017), 'Intellectual Humility: Owning Our Limitations', *Philosophy and Phenomenological Research*, 94: 509–539.

White, M. (2002), *Acid Tongues and Tranquil Dreamers: Eight Scientific Rivalries That Changed the World*. New York: Perennial.

Williams, B. (2002), *Truth and Truthfulness*. Princeton, NJ: Princeton University Press.

Yeoman, B. (2017), 'Dr. Donald Hopkins Helped Wipe Smallpox From the Planet. He Won't Rest Until He's Done the Same for Guinea Worm Disease', *Atlanta*, August. Online at www.atlantamagazine.com/health/dr-donald-hopkins-helped-wipe-smallpox-from-the-planet-now-hes-after-guinea-worm-disease.

Zagzebski, L. (1996), *Virtues of the Mind: An Inquiry Into the Nature of Virtue and the Ethical Foundations of Knowledge*. Cambridge: Cambridge University Press.

Zollman, K. (2010), 'The Epistemic Benefit of Transient Diversity', *Erkenntnis*, 72: 17–35.

10 Performance and Practice

Situating the Aesthetic Qualities of Theories

Steven French

1 Introduction

There is a considerable literature, built up over the years, which compares theories with paintings. The motivation for such a comparison is obvious: both theories, at least on the standard view, and paintings, at least those that have a depictive element, are typically taken to have a representational function (for a recent explicit consideration of this comparison see Frigg and Nguyen 2017). However, theories can also usefully be compared to other kinds of artworks such as musical compositions. Admittedly, this comparison is not so often made but it is not hard to see how it might proceed: the published 'form' (for want of a better word) of a theory corresponds to the musical score and this form is multiply reproducible, of course. Indeed, a theory, like a musical composition, can be understood as multiply instantiable. And one might go further and suggest that the presentation of a theory, at a seminar or conference for example, is akin to the performance of a musical work. Of course there are differences: the primary means by which we 'grasp' (again for want of a better word) a theory is through its publication in a journal, say, whereas for most of us, accessing a musical piece through the score is not at all straightforward. More fundamentally, perhaps, musical works might be said to *express* emotions, feelings etc. whereas even those who reject their representational function would not say the same about theories.

Nevertheless, this comparison is encouraged by Popper's placing of both theories and musical works within his framework of 'worlds': World One is the world of physical states; World Two is that of mental states and 'World Three' is the world of 'intelligibles', that is, the (crucially, objective) products of the human mind:

> Examples of world 3 objects are: the American Constitution; or Shakespeare's *The Tempest;* or his *Hamlet;* or Beethoven's Fifth Symphony; or Newton's theory of gravitation.
>
> (Popper 1978: 145)

Indeed, the significance of his reflections on the nature of music for the development of Popper's thought more generally is well documented, not least in his autobiography (1976), where he writes that it '... greatly influenced my ways of thinking in philosophy, and it ultimately led even to my distinction between world 2 and world 3' (section 13, p. 62). Both theories and musical works have aspects that may not have been immediately apparent to their author or composer respectively: 'a great work of music (like a great scientific theory) is a cosmos imposed upon chaos-in its tensions and harmonies inexhaustible even for its creator' (ibid.). Furthermore, it is by virtue of these inexhaustible tensions and harmonies that both musical works and theories may surprise us and it is this engendering of surprise that encourages us to regard both as abstract entities:

> One can, if one wishes, say that the world 3 objects themselves are *abstract* objects, and that their physical embodiments or realizations are *concrete* objects.
>
> (1978: 145)

There is more to say about the element of surprise here (see French and Vickers 2011 and French forthcoming) but I just want to emphasise the way in which the comparison of theories with musical works encourages the view of the former as in some sense 'objective'.

However, there is an obvious concern, associated with the creative process in both cases: *when*, in this process, is that objectivity achieved? Or, to put it another way, at what point in that process does the new denizen of World 3 appear (see Church 1984)? Popper himself, of course, famously reduced all such heuristic processes to psychology and explicitly sought to replace the (subjective) psychology of discovery with the (objective) logic of discovery (see again Popper 1976: section 12, p. 57), understood in the context of his overall framework of 'conjectures and refutations', but unless the Popperian were to take these World 3 entities as timeless or immortal, she needs to say something about when during the creative process they appear in that world. And preferably something that does not generate an accusation of ad-hockery.

In the next section I want to explore an alternative comparison between musical works and theories that avoids this problem, one that sees them both as products of the imagination and hence as mental entities of a sort. This too faces significant concerns but by encouraging us to look more closely at the practices associated with a theory, including presentations in seminars, conferences and the like, it suggests an alternative view that shifts the emphasis to such practices and away from both theories and musical works as objective 'things'.

2 Collingwood on Music and Theories

This comparison can be found in the reflections of Collingwood, dismissed by Popper as a proponent of the view that art is simply an expression of the artist's personality or emotions (1976: 64). That is far too quick, of course, and as we'll see, Collingwood offers a much more nuanced, but also more radical, account than Popper suggested.[1]

At the core of this account is the claim that '[i]f the making of a tune is an instance of an imaginative creation, a tune is an imaginary thing. And the same applies to a poem or a painting or any other work of art' (1938/71: 139). The role of imagination in art and science is the subject of a great deal of discussion, particularly recently (see Boden 1990; Kind and Kung 2016; Murphy, this volume), but here I just want to emphasise that Collingwood resolves the issue of how creativity is to be reconciled with the ontology of artworks by locating the latter in the same arena as the creative process, namely the imagination. However, as simply stated there's a further concern, which is that it renders artworks inter-subjectively inaccessible—in effect, there will be as many musical works labelled *Beethoven's Fifth Symphony*, say, as there are people having the relevant imaginative experiences, whether as a result of hearing a live performance or a digital download (Kania 2014).

Collingwood offers a radical response that generates an interesting comparison with science. It rests on a crucial distinction between the artist as a craftperson[2] seeking to stimulate certain emotional effects in an audience and as a 'proper' artist, aiming only to create art. In the former role, the artist produces a real thing, whether made out of stone or canvas and paint or sound. In the latter role, the artist produces something that resides only in her head, where it exists as something 'already complete and perfect' (Collingwood 1938: 139). What happens then when the piece of music is played? Then, of course, there comes into existence a 'real tune', in the sense of a sequence of noises, but as far as Collingwood is concerned this is not the work of art. As he puts it,

> [t]he noises made by the performers, and heard by the audience, are not the music at all; they are only means by which the audience, if they listen intelligently (not otherwise), can reconstruct for themselves the imaginary tune that existed in the composer's head.
>
> (ibid.)

Thus, the audience do not hear the artist's piece of music at all; instead the music, as an imaginary thing, is translated, either directly through the artist playing an instrument, say, or via a score directing some musicians' playing of instruments, into a sequence of noises, which the audience then use to reconstruct, *in their imaginations*, the artist's work. Thus the problem of subjective inaccessibility is overcome via this process of intelligent reconstruction of something imaginary.[3]

I'll leave to one side issues such as whether Collingwood's view is a form of idealism, for example, (see Ridley 1997) and how this 'preliminary account' underpins his complete philosophy of art (Davies 2008). There is a more important issue—at least as far as I'm concerned—which is that one might be sceptical that a *complete* reproduction could ever be achieved—after all, even if one accepts that feelings, emotions and so forth can be shared, one might doubt that all the nuances and complexity that were in the artist's mind during the creation of the piece could be shared in their entirety and even if they were, how would one know it? This speaks to the issue of establishing the relevant identity conditions for an artwork such as a piece of music, within the ontological framework posited by Collingwood: how can we ensure that the imaginary thing that each audience member holds is identical to the imaginary thing in the mind of the composer?

The issue of the identity conditions for artworks has, of course, a long and somewhat fractious history, which I do not intend to summarise here. Let me just say that given Collingwood's framework, it is not immediately clear how any of the favoured approaches might apply. So, for example, consider Wollheim's view (Wollheim 1968), which addresses the issue by taking a work of literature, such as *Ulysses*, or a piece of music, such as *Der Rosenkavalier*, to be types and a particular copy or performance, respectively, to be tokens of that type (ibid.: 50). The relationship between a type and its token is understood to be the most intimate of all such relations between 'generic' entities and their elements, since not only is the type present in the token, as the universal is in all its instances, but the type is often spoken about or thought of as a token itself. Of course, this naturally lends itself to the reification of the type in the form of an abstract object, and if we're not careful we find ourselves back in World 3 with Popper.

Types are then identified through consideration of the generation of the relevant tokens from a 'piece of human invention' (ibid.: 53), where the latter falls along a spectrum of cases: at one end lie poems, which come into being when certain words are written on paper, or on an iPad or whatever (or—and here Wollheim gives a nod of the head to Collingwood—when they are 'said' in the artist's head), and at the other lie symphonies, for example, which come into being when the score is written, understood as a set of instructions for a performance. All and only those properties possessed by a token of a certain type in virtue of being a token of that type are understood to be transmitted to the type, in contrast with universals (ibid.: 51).

However, setting aside all the worries one might have about taking an artwork to come into existence when a certain physical act is undertaken—worries that have to do with heuristics and creation and all the steps that have to be taken to get to the point where the poet can sit back and say they have completed their poem (see French forthcoming)—there is the crucial observation that this sort of account hardly seems to mesh with Collingwood's ontology. You might try and cash out Collingwood's distinction between a piece of music as an imaginary thing and the performance as yielding a mere tune in terms of this type-token distinction, but Wollheim's characterisation doesn't seem to be at all what Collingwood has in mind. And Wollheim takes the act of the imagination to yield a token of that type, whereas for Collingwood, of course, it is the former that yields what is to be identified with the piece of music, not the latter.

Alternatively, one could take Zemach's 'relative identity' approach (see Zemach 1991 which is a response to Taylor 1989 which in turn is a commentary on Zemach 1986), which similarly takes a piece of music to be a type, with a performance understood as an instance of that type that occurs at a particular index, where this is taken to be a triplet of a time, a place and a possible world. The identity of such pieces is then index- and type-relative, so different performances of, say Beethoven's *Fifth* must be understood as only relatively and not absolutely identical (1991: 363). This allows him to maintain that an artwork is a physical thing, without having to accept that every property of the physical thing, with which the artwork is identical, is a property of that artwork. The interpretation of identity here as absolute identity, with the consequent absurdity of that last claim, Zemach maintains, is what led Wollheim to take artworks to be abstract objects and Collingwood to take them as 'phenomenal' (1986: 241). Interpreting identity as relative avoids the problem and brings the identity criteria for artworks in line with that for things in general, in the sense that there are no identity criteria for the latter, but only for things of a certain kind, such as fall under a certain sortal term, where, he maintains, sortal terms are determined by our interests. In the case of a painting, the canvas and the artwork certainly overlap, spatio-temporally, but they have different identity conditions because they fall under different sortal terms. And the latter, as just noted, are determined by the interests served by things of that kind. For artworks, these will be aesthetic interests—so a painting such as Rembrandt's *The Night Watch*, serves the interest of being contemplated for its beauty. However, a contentious conclusion results: the identity conditions of paintings are such that paintings may exist at more than one place at the same time (p. 244). *The Night Watch*, for example, consists of different canvases, or pages in a catalogue or art book, in different places by virtue of the fact that relative to the kind that it is, serving the interests of being contemplated for its beauty, all those 'versions' of *The Night Watch* are identical—they are *The Night Watch* (again for Taylor's response and Zemach's counter, see Taylor op. cit.; Zemach 1991).

Interestingly, Zemach draws a comparison with theory change in science in order to underpin his account, writing 'We want to say that Aristotle, Newton, and Einstein disagree in describing the *same* things; ontology must be stable when ideology changes' (1991: 366). Likewise, he maintains, art critics wish to say that the same artwork may have different interpretations. But if we accept the absolute conception of identity, then we are 'edged into radical relativism' (ibid.): if the 'interpretandum' of a theory is defined by that theory then different theories will have different subject matters and no adjudication can be made between them (shades of Kuhnian incommensurability!). The way forward, he insists, is offered by relative identity, which allows us to refer to the same thing, even when there are unresolvable differences regarding the properties of that thing. And hence two construals or interpretations may still be of the same artwork, despite not being absolutely identical (ibid.: 367).

Notice, however, the shift here: Zemach is comparing scientific entities—presumably electrons, genes and the like—with paintings, rather than theories. And of course the typical realist response to his concern regarding what our friends Aristotle, Einstein and Newton can say is to appeal, first of all, to an appropriate notion of reference, that can accommodate theory change (see, for example, Psillos 1999) and, second, to maintain that there are no equivalent 'unresolvable differences' when it comes to the properties of such entities, once we accept the distinction between those properties that do the explanatory work within the theory and those that are, in effect, idle wheels (ibid.). This is quite different from the case of a painting, where one can be sure that different interpretations of *The Night Watch*, say, are about *that* painting because one can ostensively grasp what you're talking about to begin with. More importantly for our discussion, it is hard to see how this enables us to get to grips with the issue of how to establish that what I have in mind when I listen to Beethoven's *Fifth* is what you have in mind when you listen to it—to suggest that we can ostensively pick out the piece of music by, for example, pointing to members of the orchestra all sawing away would be to beg the question at issue.

Furthermore, the difference between paintings and musical works becomes apparent in this context. One might try to argue that the relationship between a musical score and a performance is akin to that between the canvas and the artwork, so that different sortal terms apply, dependent on the relevant interests in each case, but there are clear differences, not least with regard to issues concerning how the performance and artwork are composed, mereologically speaking. More productively, one could suggest that insofar as what is in the minds of the audience at a performance of Beethoven's *Fifth* serves the same interests of being listened to for its sense of 'unutterable portentous longing', say, then all such 'versions' of the music are identical. However, this only establishes the identity, relevant to that interest specific kind, of the piece of music among the members of the audience. It does not establish that the music that the audience members have in mind, reconstructed of course, is identical to that which the composer has in hers, because, of course, the interests may be entirely different—one can easily speculate that the listening experience of today's audience will most likely be shaped by Schindler's famous remark that the music signifies the struggle of a hero with fate, whereas it appears that Beethoven himself actually may have had no such motif in mind.[4]

Alternatively, one might just bite down hard on this particular bullet and accept that the work exists only in the mind of the artist and that all we can aspire to are reconstructions that may be more or less pale imitations of the work. That may seem like a counsel of despair, since it acknowledges that there can be no inter-subjective accessibility, but a Collingwoodian can insist that via appropriate listening, those reconstructions are sufficiently rich and nuanced that we can talk about, exchange opinions on, express views regarding 'the' piece—even with the artist—in such a way that we do not find ourselves talking at cross-purposes or disagreeing too often or to such a degree that we feel we are talking about different works (although that may happen). The point is, lack of such access may be no impediment to critical engagement.[5] The cost, of course, is that we are faced with the possibility of a kind of ontological proliferation once more, only this time not in some abstract space, but in the minds of all the listeners of a performance: in the mind of the composer we have 'the work', as an imaginary entity, and in the minds of all the listeners we have a plethora of more or less intelligent reconstructions, none of which are strictly or relatively identical to that work. Perhaps when it comes to musical works this can be made palatable, by, for example, emphasising the point above that it is sufficient for those reconstructions to be intelligent enough to allow the listener to grasp what it is that the composer intended to convey but let us now consider what the landscape looks like if we extend this approach to theories.

3 Comparison With Theories

So let's consider the extent to which a Collingwoodian approach might be imported into the philosophy of science. Collingwood himself compares a musical performance to a scientific presentation (1938: 140–141), noting that although they are obviously dissimilar in terms of what we are trying to 'get out of' them, there is the crucial similarity that, just as in the case of the musical performance, what we are trying to get out of the presentation is more than just the noises the speaker is making (or at least one would hope). Furthermore, the focus is not, or should not be, on the sensual pleasure one might get out of the tones of the lecturer's voice (although that might be a bonus). Rather, of course, the lecturer's intent is to develop a scientific thesis and '[t]he noises are meant to assist us in achieving what he assumes to be our purpose in coming to hear him lecture, that is, thinking this same scientific thesis for ourselves' (ibid.: 140). This is achieved not by the noises effecting the imparting of the thesis to our minds, but by their enabling us to reconstruct and thereby reproduce that thesis via 'active thinking' (ibid.).

Here too one might argue that insofar as scientific thinking is imaginative thinking, the products of that thinking—theories and models—are imaginary things. Again, worries about inter-subjective accessibility are obviated (albeit at a cost): we reconstruct and reproduce the theory concerned via this engaged and informed thinking.[6] Indeed, he insists that if we do not make efforts of the right kind, it will remain 'forever inaccessible' to us (1938: 141). One might further suggest that in this case, at least, there is less nuance to capture or that reproducing the core idea encapsulated in a scientific theory is more straightforward and thus more plausible than reproducing the emotions tied up in a piece of music, so one might be less sceptical about achieving such a reproduction.[7] And, as is clear from the quote above, Collingwood considers that what is reproduced is indeed 'the same' scientific theory, not some imitation, pale or otherwise.

Let me just note that despite his emphasis on the imagination, Collingwood's view is significantly different from the sort of fictionalism that may be familiar to philosophers of science, with its emphasis on make-believe (see for example Frigg 2010). Thus, in the build-up to his distinction between the artist as 'magician' or 'purveyor of amusement' and the 'proper' artist he notes that 'the confusion between art and amusement has been both reflected and reinforced by the confusion between imagination and make believe' (1938: 138). The latter is equivalent to imagination acting under the censorship of desire, namely the desire that the situation imagined were real, whereas the imagination required for 'proper' art is indifferent to the distinction between real and unreal. Insofar as fictionalism in the philosophy of science is likewise associated with this desire—since it is concerned with the nature and status of models that are taken to have some relationship with real situations—there seems to be a clear distinction with Collingwood's view.[8] Indeed, from the latter perspective attempts to co-opt notions of make believe from the philosophy of art into considerations of the nature of theories and models in science might be seen as inappropriately imposing elements from the justificatory phase—when one is concerned with whether one's model represents a real system or not—on the unfettered scientific imagination of the discovery or creative phase.

On Collingwood's view, then, a theory exists in the mind of the scientist concerned and we, and other scientists reconstruct that very theory in our minds when we hear it presented or read about it. But again issues must be faced as to what counts as 'the same' and more generally what would be the relevant identity conditions. Let us consider the example of Special Relativity: published in Einstein's 'annus mirabilus' (1905), it was not, of course, the work that won him the Nobel Prize. Nevertheless, no less a revered personage than Planck became a flag-bearer for the theory, giving a public lecture on it shortly after it was published and publishing his own work on it a year later, demonstrating the theory's compatibility with the Principle of Least Action (Stone 2013: 88). He also sent his assistant von Laue to Bern to meet with Einstein and the former was so impressed he published eight papers on relativity theory over the next four years. However, Planck described Einstein's theory as a *generalisation* of Lorentz's theory of electrodynamics and Einstein himself referred to it as the 'union' of Lorentz's theory and the relativity principle. Indeed, the theory was referred to as 'the Lorentz-Einstein theory', 'the relativity principle' and 'the relativity theory' over the subsequent years (Miller 1981: 88). The difference between Lorentz's theory and Einstein's has been a matter of considerable debate, of course (Darrigol 2005; Zahar 1973) with Poincare's work also thrown into the mix.

One quick and crude way of capturing the distinction is to say that whereas Lorentz's focus was on the dynamics, Einstein's was on the kinematics (and whereas Einstein rejected an absolute reference frame in his scientific work, Poincaré only did so in his philosophical reflections). It was this shift in focus that, again to put things rather crudely, underpinned the abandonment of the aether and led to the reconceptualisation of space and time (it is perhaps worth noting some recent 'pushback' in the Lorentzian direction by Brown 2007). As is well known, it was Minkowski who, also working in this general area, formulated the Minkowski space-time diagram and introduced the notions of worldline, proper time and Lorentz invariance, all now staples of relativistic discourse. Leaving aside the consideration that Einstein's 'original' 1905 theory can plausibly be said to be about clocks and rods, whereas Minkowski's formulation is about space-time (for further discussion of the latter in the context of the further distinction between 'constructive' and 'principle' theories, see Brown and Pooley 2006), Minkowski himself appears to have regarded Einstein's theory as a generalisation of Lorentz's. Einstein in return was famously dismissive of Minkowski's formulation, until, that is, he found it useful when it came to developing his 'general' theory.

Now, when Planck gave his presentation in Berlin in 1905, was he presenting the *same* theory as Einstein had published? When von Laue met Einstein in Bern and walked with him beside the river, did he have in his mind the *same* theory as Einstein did in his? More crucially, perhaps, given all the subsequent work by historians of science that has enabled us to distinguish Einstein's theory from Lorentz, if I read Einstein's 1905 paper today can I plausibly be said to have the *same* theory 'in my head' as he did; or more interestingly, perhaps, could he be said to have the *same* theory in his head then, as those of us have in our heads now, post-Minkowski?[9] The statements of Planck, Minkowski et al. and indeed Einstein himself seem to push us towards a negative answer.

The point is sharpened through consideration of what it would be to 'reconstruct' a theory in our mind. On what basis could such a 'reconstruction' take place? When I pull down my copy of A.P. French's (no relation) *Special Relativity* (French 1968), in what sense can I be said to be *reconstructing* the same theory as Einstein's, given the fundamental differences between the presentations of French in 1968 and Einstein in 1905?

This is not to say that we couldn't as a matter of principle, establish the relevant well-defined identity conditions. So, let's run Zemach's line again: in the case of Special Relativity, say, the equivalent of the canvas would be the paper of Einstein's publication and that of the artwork would be the theory and these have different identity conditions because they fall under different sortal terms.[10] And the latter are determined by the interests served by things of that kind. Thus, for the paper those interests would have to do with certain practical matters—its easily printable on, transportable etc.—whereas for the theory they would be scientific interests. A theory such as Einstein's Special Relativity serves the interests of being regarded for its empirical adequacy, say, or its truth, if one is of the realist persuasion, or its instrumental ability to generate predictions and so on. However, in this case the conclusion is not so contentious: the identity conditions of theories are such that they may exist at more than one place at the same time, which seems intuitively the case. Thus, the Special Theory of Relativity consists of different pieces of paper, or pages in a journal or textbook, found in different places, by virtue of the fact that relative to the kind that it is, serving the interests of being regarded for its truth or empirical adequacy or whatever, all those 'versions' of the theory are identical—they are The Special Theory of Relativity.

That, of course, nicely plays on a point of comparison with paintings but it would seem that Zemach's line could also be extended to the comparison that Collingwood is concerned with, namely musical works. In place of the score we have the lecturer's slides, or overheads or scribblings on the whiteboard and again the interests served by things of that kind are, in general, different from those served by theories. The latter are presumably the same for all the reconstructions effected by members of the audience, and so on that basis we should say that the audience have in mind the same theory and by extension, the same as that held by the lecturer.

This is still, of course, not necessarily the same as the theory that Einstein had in mind. Here the issue is more one of trans-temporal identity for theories and the question remains, what would the relevant conditions look like in this specific case? Answering this question requires specifying how what we take to be Einstein's theory is different from Lorentz's, Poincaré's or Minkowski's and here, as we have seen, opinions differ on the nature of that difference or whether there really is any difference to begin with. Even if we allow that reconstruction may actually take place, as a mental process, as Collingwood suggests, it remains unclear on what grounds we can confidently assert that we have, at the end of that process, the 'same' theory as Einstein had in mind. Going back to *The Night Watch*, it is as if painters through the years have 'recreated' it, using different media, different formats and so on and yet still claim it is the 'same' painting. In the case of musical works, such as Beethoven's *Fifth* where we at least have the same score, which of course may be 'interpreted' in different ways by different orchestras with different conductors, there are concerns, as we have seen, as to whether we can reasonably determine that the same musical work is being played. But in the case of theories, things are worse: we do not even have the same 'score' any more, as students at different times and in different classes are taught and physicists present and discuss, different formulations, in different media, understood in different ways.[11] Better, I suggest, to stop searching for identity conditions entirely and, consequently, to stop regarding theories as 'things' altogether.

4 Presentations as Performances

We can find similar moves in aesthetics. Ridley, for example, suggests that '. . . the whole move to ontology in thinking about music is a mistake' (Ridley 2003: 206). In part this is motivated by reflection on the actual practices of performing and listening: when we listen to Beethoven's *Fifth*, or for example, we are not concerned with establishing the ontological status of the piece, but whether it was a good or bad performance, whether, if live, it matched what we heard on a cd or piece of vinyl, or matched an earlier performance we heard, and so on (ibid.: 207). Instead of worrying about the ontology, what we should be doing, as philosophers of music, he urges, is paying (more) attention to such evaluative issues and to the *practices* of performance and listening in general.[12]

We might try to carry that last point over to presentations and seminars in science. Given the central role that such presentations play in scientific practice, there has been surprisingly little written about them within the philosophy of science. There has been some study of presentations from the perspective of science education, where it has been noted that science instruction and evaluation has typically been centred on 'the word', but that teachers 'produce other types of communicative forms during lectures, including pointing gestures, iconic gestures, diagrams, and demonstrations involving their own bodies' (Pozzer-Ardenghi and Roth 2006: 97). And to understand what an audience takes away from a lecture, we need to understand these further 'meaning-making' resources. These resources are taken to be integrated in such a way that the concept being taught is distributed across them, so that the *meaning unit* includes all such resources and their interaction in the 'meaning-making activity' of the students (ibid.: 99). The particular conception being articulated by the lecturer is then understood as revealed differently in the 'various different modalities materially enacted by speakers' (ibid.). Likewise Hwang and Roth describe lectures as 'heterogeneous performances in which meaning is synonymous with the synergistic and irreducible transactions of many different communicative modes, including gestures, body movements, body positions, prosody, and so forth' (Hwang and Roth 2011: 461). Of course, one does not have to agree with these claims about 'meaning units', and such like, to acknowledge that such resources—gestures, diagrams and so on—make an important contribution to the reception of the talk.

One can find plenty of examples of this in the history of science. Thus Peierls for example records that Planck's lectures were 'the worst I have ever attended' (1985: 19) because he simply read from the (his) textbook, whereas Sommerfeld's were a 'model of clarity' (ibid.: 23) and Caratheodory's were not very well organised but gave great insight into his thought processes (ibid.: 28–29). One could lump these comments under 'aesthetic qualities' of the presentations and regard the latter as performances. Indeed, Knight notes how, historically, lecture theatres have deserved that name and how professors became performers (Knight 2002). Davy, for example, would prepare his 'props' and rehearse with his assistants beforehand and his sense of showmanship was taken up by his successor, Faraday. There is more to say here about presentations as performances[13] and about how such 'aesthetic qualities' should not be dismissed as merely rhetorical but should be seen as significant features of the talk that help convey the meaning that the scientist intends to convey. Here again we see an obvious point of comparison with musical works and in particular with Ridley's emphasis on the practices of performance.

Of course, when it comes to theories, our evaluative moves go beyond this. As noted above, we consider whether they are empirically adequate, whether they have significant explanatory power, whether they make novel predictions and so forth. Such theoretical 'virtues' are typically regarded as distinct from aesthetic features (but see below). However, the point I want to emphasise is that however we regard them, it is their evaluative role that we, as philosophers of science, should pay attention to and not the issue of the ontological status of theories. This may then enable us to sidestep the whole debate over whether Collingwood's reconstructions are of the 'same' theory or not by insisting that *there is no theory for which identity conditions need to be provided.*

5 Theory Eliminativism

This may seem to be a radical move but eliminativist or nihilist stances are well-known elsewhere. Consider, for example, 'eliminative materialism' in the philosophy of mind (Ramsey 2016) or eliminativism about ordinary objects, for example (Korman 2016). All such views have the benefit of ontological parsimony, of course, but at the cost of having to make sense of our ordinary discourse about mental states and objects, respectively. Cameron, who advocates an eliminativist stance towards musical works as well, offers a modified form of truth-maker theory as a way of mitigating this cost (Cameron 2008a and b). The underlying intuition behind truth-maker theory is straightforward: true statements, or propositions, are 'made' true, in some sense, by certain features of reality (facts, or whatever). Opinions differ as to what this 'making' consists of—whether it has to do with entailment, or primitive necessitation or some kind of 'grounding' (see MacBride 2014)—and both what the truth-makers are (facts, features, whatever) and what the truth-bearers are (statements, judgements, propositions, etc.). Given the context, I shall adopt the minimal assumption that the truth-bearers are *representational* (ibid.) and that the truth-makers are what are deemed to be certain *fundamental features* within the relevant discourse. So, we may say that the statement 'the electron has charge *e*' is made true by those fundamental features of the world having to do with charge.

Now consider the statement 'electrons exist'. What makes this true? Standardly, we would take the relevant truth-maker to be the set of elementary particles that we label 'electrons'. In general, on the standard understanding of truth-maker theory, the truth-maker for the claim '*x* exists' is always *x* (see, for example, Armstrong 2004). However, Cameron allows for the truth-maker of the sentence to be something other than '*x*':

> I think one of the benefits of truthmaker theory is to allow that '*x* exists' might be made true by something other than *x*, and hence that '*a* exists' might be true according to some theory without *a* being an ontological commitment of that theory.
>
> (Cameron 2008b: 4)

To see how this works, let's take that much used and abused example of the statue (e.g. Michelangelo's *David*). Now, take the statement, 'there are statues'. In terms of what there is at the most fundamental level of the world, there are no statues, only elementary particles, arranged in a certain way. However, we may still accept that the statement, as expressed in everyday English, 'there are statues' is true, *but* not in virtue of the fact that there are statues, since the eliminativist denies this, but rather in virtue of the fact that there are elements of our fundamental ontology that are arranged in a certain way (Cameron 2008a: 301). Note how this device allows us to be eliminativists about, in this case, statues without having to back away from the truth of all the claims and statements we typically make in everyday English. We can continue to maintain that 'There is a statue' is true, without having to admit statues into our metaphysics, because what makes that statement true is not some element of our ontology called a statue, but rather certain features of the world—whether elementary particles arranged a certain way or whatever (for further on how this approach can be extended to the objects of both the 'everyday' and of physics in a structuralist context, see French 2014).

Cameron deploys this device to avoid commitment to the existence of musical works as abstract objects (Cameron 2008b) and we can also apply it to the case of theories (French and Vickers 2011 see also Vickers 2014). Thus we can eliminate theories from our ontology, yet still talk about them, their inter-relationships, empirical adequacy, simplicity, indeed their various qualities and so forth. What makes this talk true (or false) are not the entities directly referred to in it (namely the theories), but rather certain features that exist at some fundamental level, to which we *are* ontologically committed. Thus we retain all the advantages of being able to talk about theories at the everyday level, as it were, but we avoid having to inflate our ontology unnecessarily.

The question now, of course, is what are the truth-makers in our discourse about theories. There are two obvious candidates: the *thoughts* of individuals in the relevant scientific community and scientific *practices*.

Let us consider each of these options in turn.

6 Thoughts[14]

Here there are obvious connections to Collingwood's account. So, we could begin by saying that what matters vis-à-vis theory-talk are the mental *representations* (about a certain domain of phenomena) common to individuals in the relevant scientific community. The difficulties sketched above now do not arise, since we are no longer *identifying* theories with these thoughts: fundamentally speaking, theories do not exist. So we don't have to follow Collingwood in regarding theories, or artworks for that matter, *as* ideal, mental entities, in whatever sense; nor do we have to fret over whether in listening to a scientific presentation or reading a textbook we are reproducing or reconstructing the same theory or a kind of simulacrum. And of course, we don't have to worry about the identity conditions for theories in general because, again, from this eliminativist perspective, there are none!

But theory-*talk* can be assessed by looking to what is going on in the minds of those in the relevant scientific community. One can then take statements such as, 'Special Relativity is based on two postulates (the principle of relativity and the constancy of the speed of light)' as made true by an understanding of that statement and its component elements shared by relevant scientists. Of course it may be that this statement turns out *not* to be accepted by everybody in the relevant scientific community, or even that the contrary claim is accepted. So, some formulations invoke just one postulate, namely that of Lorentz covariance. The relevant cognitive facts may then in turn support claims to the effect that 'Special Relativity is based on only one postulate'. The issue as to whether Special Relativity, as a theory, should be considered broad enough to encompass two postulate and one postulate formulations or whether the introduction of forms of the latter should be regarded as introducing an entirely new theory takes us back into the thicket of issues concerning the relevant identity conditions and again, with one bound eliminativism sidesteps this issue.

Of course, this line raises a number of questions. How exactly should we understand these 'mental representations' that do the truth-making work? What is the relationship between a mental representation in one mind and that in another? Can these ever truly be exactly the same? If not, how should we think about the *similarity* of one mental representation to another? These questions lead into variants of the concerns raised with regard to Collingwood's stance on artworks: if you and I have different relevant mental representations, can they both be taken as truth-makers of statements such as the above? Or should they be taken as exemplifications of the same truth-maker? In which case, how similar do they have to be for that to be the case? Or should they be taken as different truth-makers entirely, so that there may be as many truth-makers for the statement as there are minds thinking about it?

Perhaps it might seem plausible to say that in such cases there is only one truth-maker for a statement like 'Special Relativity is based on two postulates', namely a complicated network of overlapping mental representations in the minds of individuals in the relevant scientific (sub-) community. This might help assuage the concern that given how complicated theories can be, and how they involve abstract ideas, and, often, complicated mathematics, nobody can hold such a scientific theory in their own head. Alternatively, we could follow Giere (2002), who argues that even multiplying together 456 and 789 usually means 'creating and manipulating *external representations*' (ibid.: 288, original emphasis). In other words, we have to start writing things down to work through the problem. Similarly, it might be claimed, theories cannot be conceived unless we start writing things down to assist our minds.[15] Giere goes on to recommend that we think in terms of *distributed cognitive systems*, essentially a combination of the internal (whether in a single individual or in many) and the external. We could then adapt this to suggest that only those truth-makers that relate to fragments of a theory exist within any one mind at any one time.

But of course as truth-makers, these mental representations are still inter-subjectively inaccessible—or at least, they are as long as we keep the resultant speech, behaviour and, crucially, *practices* out of the picture. But, including those invites a move to a very different picture, as we'll now see.

7 Practices

The alternative option that I want to explore takes the relevant truth-makers—that is, all that *really* exists in this context—to be the complex set of practices of the scientific community: the writing and dissemination of articles, the performance of experiments and of lectures, and so on. These are all either concrete entities or reducible to such and within this particular context, ontological commitment to such practices can be taken at face value. These practices will then collectively act as the truth-makers for our theory talk.

So, take the claim that 'Special Relativity is empirically adequate'. This can be understood in terms of the embedding of empirical sub-structures into theoretical structures, but it is not the embedding itself that makes true that claim, since that would presuppose there is an entity—the theory—into whose structure the empirical sub-structures are embedded. Rather, it is made true by a complex nexus of practices involving the obtaining of predictions or the identification of experimental consequences more generally, the testing of such consequences through experiments and so forth. These practices result in datasets, for example, and underpin models of that data, and of the experiment, and so on, as Suppes famously described and all of these, or the nexus they constitute, can then be represented (by *us*, philosophers of science) set-theoretically as empirical sub-structures. However, when it comes to what functions as the relevant truth-maker, we should not confuse that representation with the practices themselves.

Now, consider again the statement 'Special Relativity is based on two postulates'. Again, to determine if the claim is true, we draw on certain practices. These can be quite minimal in this case, as in simply opening a (trusted) textbook, noting the relevant formalism and identifying the two postulates. Alternatively, we might reflect on those claims that Special Relativity only needs one postulate and we might perhaps attempt to demonstrate the truth of the claim ourselves, working through any mathematics involved, where such 'working through' itself constitutes a form of practice, of course. And we can also extend this approach to the aesthetic qualities of theories.

8 Practices and Aesthetic Virtues

So, consider now the claim 'Special Relativity is an elegant theory'. Scientists themselves often make statements such as this that appear to assign aesthetic values to theories (for a useful selection of quotes from scientists extolling such values, see Ivanova 2017). Furthermore, it has been argued that such features as beauty and elegance play a role in not only the 'discovery' of a theory—that is, as a heuristic factor—but also with regard to the *justification* for choosing the theory (Ivanova 2017). Such claims are contentious, not least because they demand explication with regard to the purported relationship between beauty or elegance and truth (ibid.).[16] Here I want to focus on the implication that statements such as the above seem to refer directly to features of the theory *as an entity*.

How are similar claims about artworks treated in the philosophy of art? Setting things out rather crudely, there is a well-known split between those who argue that aesthetic qualities such as 'elegance' supervene on non-aesthetic properties of the artwork and those who deny this and argue, for example, that context pays an important role in the assignment of such qualities (for a useful overview of the debate, see Hick 2012). Consider, for example, Turner's painting 'Thomson's Aeolian Harp' (www.tate.org.uk/art/artworks/turner-thomsons-aeolian-harp-tw0579), described as a 'truly elegant painting' (Ziff 1980). On the one hand, the property of elegance might be taken to supervene on various features of the painting, both structural and non-structural, such as the balance of shade and light between the land and the sky, the sweep of the river in the centre, which draws the eye to the figures on the right and so on.

On the other hand, it could be argued that such an ascription can only be made in a certain context—in this case that which embraces our current aesthetic sensibilities and our understanding of Turner and *his* context. Present the painting to a devotee of the Bauhaus school and they might twist Constable's praise of Claude (who inspired Turner at this stage of his career) and dismiss it as entirely too 'amiable'. Still, we could fold such contextual or non-intrinsic features into the supervenient base and still maintain that the quality of elegance derives, at least in part, from features of the painting, as an art object.[17] In either case, it is the object itself that is said to possess the relevant aesthetic quality. And we can extend this analysis to musical works, of course, although here one might want to pay attention to the distinction between attributing elegance, say, to the work itself—conceived as an abstract artefact or whatever—and the performance. Thus one of Bach's toccatas might be described as 'having a tendency to ramble' yet be considered to have been played elegantly by a certain pianist (see Distler 2015).

It would seem that we could reproduce these sorts of moves when it comes to scientific theories and models. So, take the Crick and Watson model of DNA. In this case the physical entity might also be said to possess a certain elegance, in exactly the same aesthetically informed sense as a sculpture, say. (It might also be said to be elegant in a scientifically informed sense, but I'll come to that shortly.) And we can obviously extend this to scientific drawings, pictures, diagrams and so forth; consider the 'Beautiful Science' exhibition held at the British Library, for example (www.bl.uk/reshelp/experthelp/science/inspiringscience/2014/beautifulscience.html)—who could deny that NASA's visualisation of the ocean currents is beautiful and aesthetically pleasing?

But what about non-physical models and theories? Take the example of Newtonian mechanics: it might well be claimed that this is an elegant theory, where this quality can be attributed to the internal structure of the theory,[18] or its simplicity, or some combination of these and other features (just as with Turner's painting); or it might be insisted that such an attribution is contextual—so, some might argue that Newton's original geometric presentation of his theory in the *Principia* is not elegant at all (see: http://cudl.lib.cam.ac.uk/view/PR-ADV-B-00039-00002/1) and that it only acquires this quality when formulated in the modern notation. Relatedly, it is well known that expressing Newton's equations in polar coordinates, say, can be clumsy and tedious, and that the Lagrangian formulation, which is independent of the coordinates, offers a particularly elegant presentation of the theory.

Exporting such moves from aesthetics to the philosophy of science may suggest that the truth-maker of our claim 'Special Relativity is an elegant theory' is the quality of elegance as possessed by the theory *qua* object, just as in the case of the Turner painting. But this is to move too fast. First of all, note that in the (non-physical) examples above we are talking about different formulations and presentations; indeed, the attempt to construct a parallel form of contextualisation makes this clear. So one can press this point and insist that whatever quality scientific elegance is or consists in, it is attributed not to 'the theory', as an object, whether abstract or not, but to a set of symbols laid down in a certain order, whether on paper or a whiteboard or whatever, which are then manipulated in a certain way in practice. And to infer from this that such qualities can also be attributed to a theory, because such formulations are *of* the latter, is to beg the question, of course! Thus we can argue that all that the above examples show is that a set of symbols in a certain order possesses aesthetic 'elegance', say, and that insofar as this is what we mean when we utter the above assertion, then it should be taken as shorthand for 'Special Relativity as expressed via a certain set of symbols is an elegant theory'. And it is no objection to eliminativism to say that the truth-maker of 'That expression of Newton's theory is elegant' is a certain aesthetic quality possessed by a set of symbols set down in a particular order.

If one took the line that the relevant truth-makers are thoughts, then one can understand the truth of statements such as, 'Special Relativity is an elegant theory' as given in terms of some aesthetic element associated with the relevant mental phenomena. That phenomena might be regarded as itself irreducibly aesthetic, in the sense that there is a particular such phenomenon associated with the appreciation of things—artworks, theories, whatever—that are regarded as elegant; or it might be regarded as non-aesthetic and associated with a feature, such as symmetry say, that in general is not always associated with elegance but in the case of theories, and artworks perhaps, it is.

Alternatively, we might insist either that it is downright inappropriate to attribute aesthetic qualities to scientific theories and models or, at the very least, that what is typically meant by 'elegant' here is not what is meant in the case of paintings or musical works. So, a typical way of cashing this out in the scientific context is via a combination of parsimony and power, for example (for an overview of such reductive moves, see Ivanova 2017).[19] With regard to the former, this can be explicated in terms of *qualitative* parsimony, in the sense that a theory with fewer *kinds* of entities might be preferred over its rivals and *quantitative* parsimony, in which preference is given to the theory that minimises the *number* of entities postulated. Not only are such preferences evidenced in the history of science (classic examples that are used include the postulation of the neutrino in beta-decay, Avogadro's laws and the positing of the existence of Neptune), but epistemic justification can be given for them: for non-parsimonious theories to appropriately entail the evidence, certain additional costly (and context dependent) assumptions must be introduced that their parsimonious competitors can avoid (Jansson and Tallant 2017). Thus, the truth-maker of 'Theory *T* is more parsimonious than theory *T**' will be found in the practices we undertake in showing how the former is less 'costly' in the sense of being less complex, having fewer additional assumptions etc than the latter.

What about power? This can be understood in various ways but typically as unificatory or explanatory.[20] Thus, we could regard the theory that expresses or includes the conservation of momentum for all systems as more powerful and hence as more 'elegant' than a theory that incorporates conservation of momentum for elastic processes only (Post 1960: 35). In such cases, the truth-maker for 'Theory *T* is more powerful than theory *T**' can be associated with the practices we engage in when we (apparently) apply *T* and *T** to phenomena, where the range of phenomena will be wider in *T*'s case and hence the number and kinds of practices (theoretical an experimental) will be different. Likewise when it comes to explanatory power—insofar as *T* offers a better explanation than *T** that will be manifested in the different practices involved in each case (involving fewer, simpler etc. deductions, for example, on the Deductive-Nomological view of explanation, or featuring fewer, again, or different difference makers and so on).

The upshot then is as follows: if what we mean by 'Special Relativity is an elegant theory' is just that a certain expression 'of' the theory possesses certain aesthetic qualities, then insofar as that expression is not the theory, nor is it *of* the theory in any metaphysical serious sense, then this is straightforwardly accommodated. But if we mean something scientifically 'deeper' by elegance, then insofar as that can be understood in terms of some combination of parsimony and power, the statement is made true by the relevant practices, involving, for example, the ease of deduction of certain (written, typed, scrawled and so on) statements from the axioms or fundamental claims of the theory, the way in which a wide variety of claims (both theoretical and empirical) can be obtained from these axioms and so on (the details of which aspects of practice will act as truth-makers will of course depend on precisely how one understands both parsimony and power, as the different accounts of explanation illustrate).[21]

9 Conclusion

Leaving aside his distinction between craft and art and so forth, Collingwood was on to something when he emphasised the reconstructive nature of artistic experience. When we read a scientific text or paper or attend a seminar, we are indeed engaging with that text or talk in a way that may appear as if we were reconstructing 'the' theory. But that appearance comes freighted with dubious ontology. Instead, I would suggest, we are engaging in certain practices that supply the truth-makers for certain claims that are putatively 'about' theories but in fact all there *is*, are these practices. Realising that may free us up to consider various aspects of the history and philosophy in a new way. In particular, consider theory change and the way this is usually portrayed as driven by certain 'virtues' taken to be 'of' theories, with the implicit understanding of the latter as entities of some kind. Dropping that understanding and shifting our attention to the diverse practices that make true statements referring to such virtues—whether 'theoretical' or 'aesthetic'—opens the door to a more fine-grained and nuanced appreciation of the factors that are involved when the commitments of the relevant scientific community shift.

Consider for example the recent debate on one of the central topics of this collection, occasioned by the publication of Hossenfelder (2018). There it is argued that current physics has taken a wrong turn and finds itself in a kind of disciplinary cul-de-sac by virtue of an unwarranted emphasis on looking for theories that have the virtues of 'beauty' and 'naturalness'. As Hossenfelder argues and as Butterfield (forthcoming) also notes, what is taken as 'beautiful' or 'natural' and how much weight one should give to these 'virtues', may vary across history, across disciplines and between scientists. But this only counts against their purported role in theorising if it is assumed they must be 'objective' in the sense of applying to or being instantiated in, theories understood as entities. The alternative would be that they are 'merely' subjective and hence immediately dismissable. However, we can perhaps find a middle road between these alternatives if we shift our epistemic focus from the theories themselves to the practices that make true claims involving such virtues. This will also reveal that scientists' preference, or not, for such 'virtues' is less a matter of personal foible and more to do with particular practices and, delving even deeper, their form, their history, their connections to other such practices and so on.[22] It may also help us better understand why, when other virtues such as empirical adequacy or predictive power are out of reach, these come to the fore in scientists' reasoning.[23] Indeed, the whole debate of 'Why trust a theory?' (the title of a recent conference; see Dardashti, Dawid and Thebault 2019 for the proceedings, with that title) given the current invocation of such virtues, can perhaps be regarded as mis-framed. Re-orienting it in terms of the relevant practices may result in a more fruitful approach to the underlying concerns.

Acknowledgements

Many thanks to Milena Ivanova and Alice Murphy for useful discussions and comments and, albeit in the distant past, to Pete Vickers for our joint work on the view that there really are no things that are theories.

Notes

1. A useful collection of papers on Collingwood's philosophy can be found in 'Collingwood and the Philosophy of History', *Journal of the Philosophy of History* 11 (2017).
2. Or as Collingwood characterises her/him, as a magician or purveyor of amusement (Collingwood 1938: 139).
3. There is, then, an obvious element of elitism in this view, since to obtain such an imaginative experience, one must be 'properly qualified' (ibid.: 148). As we'll see that quite straightforwardly applies to theories.
4. But perhaps this is to confuse what goes on in creation with what the artist ultimately intends the piece to convey and if one can 'get' that, one could legitimately be said to 'have' the musical work in one's mind; cf. Davies *op. cit.*, pp. 170–171.
5. Interestingly, Collingwood also argues that '[t]he work of artistic creation is not a work performed in any exclusive or complete fashion in the mind of the person whom we call the artist' (1938: 323–324). To suppose otherwise is to fall prey to the delusion of individualistic psychology. Instead, aesthetic activity should be thought of as a 'corporate' activity belonging to a community consisting of not only the artist but all those who have influenced her, together with the performers, who, in the case of a musical work, say, effectively collaborate with the artist and, finally, the audience, which also collaborates in the creation of the work. It is not clear to me how one might reconcile this view of art with the claim that artworks are imaginary. Collingwood himself insists that this 'corporate' stance is not inconsistent with the view that aesthetic activity goes on in the artist's mind, because for a speaker to have the experience of being listened to there must be a listener, but that doesn't quite address the issue. One could insist that it is this latter stance that represents the more mature account, with the first relegated to preliminary musings (see Davies *op. cit.*), but that also does not do complete justice to the claims noted above. There is also the even more radical stance that a 'tune' is indeed an imaginary thing but the imagination involved is that of the community as a collective!
6. As with musical works, it is only those who are appropriately scientifically informed, through their training say, who are capable of such reconstruction.

7. Of course, the sceptic might press the point that a major scientific theory, at least, such as *General Relativity*, or quantum mechanics, is just as nuanced and complex in its implications as Beethoven's *Fifth*, say, and, furthermore, one might well have reasonable doubts as to whether anyone can reproduce what was in Einstein's mind, say, no matter how well trained or appropriately educated!
8. There is more to say, of course, but especially with regard to the distinction between the 'real' and the 'unreal' and how that maps—or not—onto that between 'fact' and 'fiction'. Friend, for example, has argued that 'works of non-fiction may invite the same imaginative responses as fiction, just as works of fiction may invite the same cognitive responses as non-fiction' (2008: 151). However, what Collingwood seems to have been concerned about are the constraints that the desire to establish a relationship with the real introduce and distinguish make-believe from imagination *per se*. I'd like to thank Alice Murphy for pressing me on this.
9. And also post-conventionalism, post-debate over substantivalism.
10. Whether they could be said to overlap, as the canvas of a painting and the artwork do, is a further issue—not spatio-temporally, of course, but perhaps one could expand the notion of 'overlap' in some metaphysical sense.
11. You might insist that the original published version of the theory is the equivalent of the score, but leaving aside issues as to why the former should be accorded such significance, French (the other one) does not call his presentation of Einstein's theory an 'interpretation' nor do any of the other physics teachers who present it via books, lectures, YouTube videos or whatever.
12. As Milena Ivanova has suggested, such evaluation opens the door to biases of various forms. In science, for example, gender stereotypes are known to influence the evaluation of not only presentations, but also, at least in certain cases, the publication of results.
13. And not just with regard to the early history of scientific lectures as this sort of analysis could be extended to YouTube presentations or even scientists' 'performances' on social media.
14. Here I'm relying on the input of Pete Vickers as presented in French and Vickers (2011).
15. And here one might draw on some of the arguments associated with 'extended cognition', for example.
16. In this regard Ivanova (ibid.) makes an intriguing suggestion that exports certain results at the intersection of aesthetics and psychology into the philosophy of science: certain studies have shown that exposure to so-called 'bad' art is not correlated with an increase in the subjects' aesthetic appreciation, leading to the conclusion that some feature over and above such exposure must be responsible for the subject's aesthetic responses to artworks (Meskin, Phelan, Moore and Kieran 2013). She suggests that if such results could be reproduced with regard to scientific theories (and models), we would have grounds for similarly concluding that exposure and habituation cannot solely be responsible for scientists' appreciation of such aesthetic features.
17. Hick himself argues for the view that 'an object's aesthetic properties just *are* its realized powers to produce aesthetic effects of particular kinds in suitable perceivers under suitable conditions' (Hick 2012: 314). Leaving aside the issue of whether artworks can really be said to possess 'powers', in a metaphysically robust sense, as Hick notes this raises questions about which perceivers? And under what conditions? Interestingly, however, he acknowledges that scientific and mathematical formulae may be accredited with aesthetic properties and, as with everyday or natural objects, also so accredited,

it would obviously be inappropriate to regard such properties as dependent on certain artistic categorisations or art-history contextualisation.

18. There may be a distinction to be made between the attribution of aesthetic features to the relevant *mathematics* and (purportedly) to the theory 'as a whole' (for examples of aesthetic responses to mathematics see again Ivanova 2017).
19. Cf. Post's distinction in a much-neglected paper, between 'linguistic' and 'semantic' simplicity (Post 1960), where the former refers to the kind of simplicity in or elegance of formulation already discussed, and the latter covers unificatory and explanatory power.
20. And on the view that explanation fundamentally consists in unification then the latter will just collapse into the former.
21. Just to emphasise: with all these statements, I am not talking about the truth or falsity of claims the theory makes about the *world*, but rather the truth or falsity of claims made about the *theory*.
22. It may help us understand how Dirac, for example, who is often held up as a representative of those who extol the virtue of beauty (Butterfield forthcoming) could also play so fast and loose with the relevant mathematics (as with the 'delta function') and adopt what might broadly be described as an 'engineering perspective'.
23. There is much more to say of course, not least about the relationship between such virtues and truth; for a discussion of how eliminativism and realism may co-exist, see French (2017).

Bibliography

Armstrong, D. M., (2004), *Truth and Truthmakers*, Cambridge: Cambridge University Press.

Boden, M. (1990), *The Creative Mind: Myths and Mechanisms*. Weidenfeld & Nicholson.

Brown, H. R. (2007), *Physical Relativity: Space-time Structure From a Dynamical Perspective*. Oxford University Press.

Brown, H. R. and Pooley, O. (2006), 'Minkowski Space-Time: A Glorious Non-Entity', *Philosophy and Foundations of Physics*, 1: 67–89.

Butterfield, J. (forthcoming), 'Review of Hossenfelder, S. (2018). *Lost in Math: How Beauty Leads Physics Astray*, Basic Books', *Physics in Perspective*.

Cameron, R. (2008a), 'There Are No Things That Are Musical Works', *British Journal of Aesthetics*, 48: 295–314.

Cameron, R. (2008b), 'Truthmakers and Ontological Commitment', *Philosophical Studies*, 140: 1–18.

Church, R. (1984), 'Popper's World 3 and the Problem of the Printed Line', *Australasian Journal of Philosophy*, 62: 378–391.

Collingwood, R. G. (1938/71), *The Principles of Art*. Clarendon Press.

Dardashti, R., Dawid, R. and Thebault, K. (2019), *Why Trust a Theory: Epistemology of Fundamental Physics*. Cambridge University Press.

Darrigol, O. (2005), 'The Genesis of the Theory of Relativity', *Séminaire Poincaré*, 1: 1–22.

Davies, D. (2008), 'Collingwood's "Performance" Theory of Art', *British Journal of Aesthetics*, 48: 162–174.

Distler, J. (2015), 'Elegant Bach Toccatas', *Classics Today*. Online at www.classicstoday.com/review/elegant-bach-toccatas/

Einstein, Albert. 1905. "Zur Elektrodynamik bewegter Korper", *Annalen der Physik*, 17: 891–921; English trans.: 'On the Electrodynamics of Moving Bodies', Translation by G. B. Jeffery and W. Perrett in *The Principle of Relativity*, London: Methuen and Company, Ltd. (1923).

French, A. P. (1968), *Special Relativity*. MIT Press.

French, S. (2014), *The Structure of the World*. Oxford University Press.

French, S. (2017), '(Structural) Realism and Its Representational Vehicles', *Synthese*, 194: 3311–3326.

French, S. (forthcoming), *There Are No Such Things as Theories*. Oxford University Press.

French, S. and Vickers, P. (2011), 'Are There No Things That Are Theories?', *The British Journal for the Philosophy of Science*, 62: 771–804.

Friend, S. (2008), 'Imagining Fact and Fiction', in K. Stock and K. Thomson-Jones (eds.), *New Waves in Aesthetics*. Springer, pp. 150–169.

Frigg, R. (2010), 'Models and Fiction', *Synthese*, 172: 251–268.

Frigg, R. and Nguyen, J. (2017), 'Of Barrels and Pipes: Representation-as in Art and Science', in O. Bueno, G. Darby, S. French and D. Rickles (eds.), *Thinking About Science, Reflecting on Art*. Routledge, pp. 41–61.

Giere, R. (2002), 'Scientific Cognition as Distributed Cognition', in P. Carruthers, S. Stich and M. Siegal (eds.), *The Cognitive Basis of Science*. Cambridge University Press, pp. 285–299.

Hick, D. H. (2012), 'Aesthetic Supervenience Revisited', *British Journal of Aesthetics*, 52: 301–316.

Hossenfelder, S. (2018), *Lost in Math: How Beauty Leads Physics Astray*. Basic Books.

Hwang, S. W. and Roth, W.-M. (2011), 'The (Embodied) Performance of Physics Concepts in Lectures', *Research in Science Education*, 41: 461–477.

Ivanova, M. (2017), 'Aesthetic Values in Science', *Philosophy Compass*, 12: e12433. https://doi.org/10.1111/phc3.12433.

Jansson, L. and Tallant, J. (2017), 'Quantitative Parsimony: Probably for the Better', *British Journal for the Philosophy of Science*, 68: 781–803.

Kania, A. (2014), 'The Philosophy of Music', in E. N. Zalta (ed.), *The Stanford Encyclopedia of Philosophy*, Spring 2014 ed. Online at http://plato.stanford.edu/archives/spr2014/entries/music/.

Kind, A. and Kung, P. (2016), *Knowledge Through Imagination*. Oxford University Press.

Knight, D. (2002), 'Scientific Lectures: A History of Performance', *Interdisciplinary Science Reviews*, 27: 217–224.

Korman, D. Z. (2016), 'Ordinary Objects', in E. N. Zalta (ed.), *The Stanford Encyclopedia of Philosophy*, Spring 2016 ed. Online at https://plato.stanford.edu/archives/spr2016/entries/ordinary-objects/.

Macbride, F. (2014), 'Truthmakers', in E. N. Zalta (ed.), *The Stanford Encyclopedia of Philosophy*, Spring 2014 ed. Online at http://plato.stanford.edu/archives/spr2014/entries/truthmakers/.

Meskin, A., Phelan, M., Moore, M. and Kieran, M. (2013), 'Mere Exposure to Bad Art', *British Journal of Aesthetics*, 53: 139–164.

Miller, A. (1981), *Albert Einstein's Special Theory of Relativity: Emergence (1905) and Early Interpretation (1905–1911)*. Reading: Addison-Wesley.

Murphy, A. (2020), 'The Literary Form of Scientific Thought Experiments', in M. Ivanova and S. French (eds.), *Aesthetics of Science*. Routledge.

Peierls, R. (1985), *Bird of Passage: Recollections of a Physicist*. Princeton University Press.

Popper, K. (1976), *Unended Quest: An Intellectual Autobiography*. Routledge.

Popper, K. (1978), *Three Worlds: The Tanner Lectures on Human Values*. University of Utah Press.

Post, H. R. (1960), 'Simplicity in Scientific Theories', *British Journal for the Philosophy of Science*, 11: 32–41.

Pozzer-Ardenghi, L. and Roth, W.-M. (2006), 'On Performing Concepts During Science Lectures', *Science Education*, 91: 96–114. https://doi.org/10.1002/sce.20172.

Psillos, S. (1999), *Scientific Realism: How Science Tracks Truth*. Routledge.

Ramsey, W. (2016), 'Eliminative Materialism', in E. N. Zalta (ed.), *The Stanford Encyclopedia of Philosophy*, Winter 2016 ed. Online at https://plato.stanford.edu/archives/win2016/entries/materialism-eliminative/.

Ridley, A. (1997), 'Not Ideal: Collingwood's Expression Theory', *Journal of Aesthetics and Art Criticism*, 55: 263–272.

Ridley, A. (2003), 'Against Musical Ontology', *Journal of Philosophy*, 100: 203–220.

Stone, D. (2013), *Einstein and the Quantum: The Quest of the Valiant Swabian*. Princeton University Press.

Taylor, P. (1989), 'Painting and Identity', *British Journal of Aesthetics*, 29: 352–362.

Vickers, P. (2014), 'Scientific Theory Eliminativism', *Erkenntnis*, 79: 111–126.

Wollheim, R. (1968), *Arts and Its Objects*. Harper Row.

Zahar, E. (1973), 'Why Did Einstein's Programme Supersede Lorentz's?' (I). *British Journal for the Philosophy of Science*, 24: 95–123; and (II). 223–262.

Zemach, E. (1986), 'No Identification Without Evaluation', *British Journal of Aesthetics*, 26: 239–251.

Zemach, E. (1991), 'Art and Identity', *British Journal of Aesthetics*, 31: 363–368.

Ziff, J. (1980), 'Review of *The Paintings of J. M. W. Turner* by Martin Butlin, Evelyn Joll', *The Art Bulletin*, 62: 166–171.

Contributors

Arcangeli, Margherita, Post-doctoral fellow, Institut Jean-Nicod

Bird, Alexander, Peter Sowerby Professor of Philosophy and Medicine, Department of Philosophy, King's College London

Dokic, Jérôme, Professor, EHESS and Institut Jean-Nicod

Elgin, Catherine Z., Professor of Philosophy, Harvard University

French, Steven, Professor of Philosophy of Science, University of Leeds

Ivanova, Milena, Teaching Associate in the Department for History and Philosophy of Science, University of Cambridge and Bye-Fellow at Fitzwilliam College

Kieran, Matthew, Professor of Philosophy and the Arts, University of Leeds

Meynell, Letitia, Associate Professor of Philosophy, cross appointed with the Gender and Women's Studies Programme, at Dalhousie University in Nova Scotia, Canada

Murphy, Alice, Postgraduate Researcher, University of Leeds

Todd, Cain, Senior Lecturer in Philosophy, Lancaster University

Name Index

Subject Index

For Product Safety Concerns and Information please contact our EU representative GPSR@taylorandfrancis.com
Taylor & Francis Verlag GmbH, Kaufingerstraße 24, 80331 München, Germany

www.ingramcontent.com/pod-product-compliance
Lightning Source LLC
LaVergne TN
LVHW011128300926
849160LV00008B/567
9781032337180